AuthorHouse™
1663 Liberty Drive
Bloomington, IN 47403
www.authorhouse.com
Phone: 1 (800) 839-8640

Published by AuthorHouse 07/13/2016

ISBN: 978-1-5246-1846-9 (sc)
978-1-5246-1847-6 (e)

Library of Congress Control Number: 2016911236

Print information available on the last page.

Any people depicted in stock imagery provided by Thinkstock are models,
and such images are being used for illustrative purposes only.
Certain stock imagery © Thinkstock.

This book is printed on acid-free paper.

Because of the dynamic nature of the Internet, any web addresses or links contained in this book may have changed
since publication and may no longer be valid. The views expressed in this work are solely those of the author and do
not necessarily reflect the views of the publisher, and the publisher hereby disclaims any responsibility for them.

authorHOUSE®

The First and Original Inventor

Human Powered Transportation Means
for the 21st Century

Human Powered Helicopter

Dragster

By Richard Chastain M.E.T.

Volume 1

A Cut Above Mechanical Engineering Technology
and Short Stories

Index

Introduction

Be ready to spend the next however-long copying off of my research. This publication is being made to void the obviousness to prevent patent for these ideas and discoveries. It is with my every intention to secure the post of First and Original Inventor of these ideas and discoveries, and I am publishing my work to affect no one else applying for patent or no one being not able to. Crowdfunding websites have attracted no interested contributors to my projects, and attempting to secure sponsorship funding may only yield first filing competition for the patent. Either this publication will make it so no one can file a patent application, or anyone will be able to and that will make it so I can, and without the required arithmetic which is included with this publication all other attempts to file patent applications will probably be by trial and error and my developed products will always be more powerful and more accurately designed, and if a filed application for patent uses my mathematicals to secure the accuracy and output for "their" application of "first and original inventor", use of my copyrighted work will be involved and there may be further dealings to contend with with these "inventors", and I still have to make a soft copy of my Laboratory Notebook, and that will really screw things up for anyone filing a patent application and claiming to be first to file when signing the application by anyone other than me after I have produced a publication and established that I am the First and Original Inventor should be inconsistent with lawful activity. What will be with the signing requirements and being able to sign a patent application after this publication is published? Who is at fault here, and since when did an author need a claim number for algebra, geometry, trigonometry, basic functions; square root of 2, base powers, and other elements in the public domain. I didn't know the ancient Egyptians still had copyrights.

There may be some repetition of material in the publication but it is all just good practice. I hope you enjoy rifling through my work. As it may be possible to realize, this is not just some work which may be "talked" about, and illegal lines of questioning are virtually impossible to respond to. You might say "Explain yourself!" is insanity. This is the end result of what started in November 1988 and all there is here is the conclusion of all of the working out of all the "bugs" and the perfecting of it all. This work constitutes virtually everything I wanted to know about everything I ever wanted to know anything about, and I still have more to ascertain within the means of my capability although to me it is routine now.

What fraction of the radius (pi r² (r/n)2 do I **divide the area** the radius by to get the length of the radius I multiply to the area of the diameter to equal the volume of the sphere?   1.5 = 2/3

———multiply

pi r² (r/(r/(4/3 pi r³ /pi r² 2)) = volume.

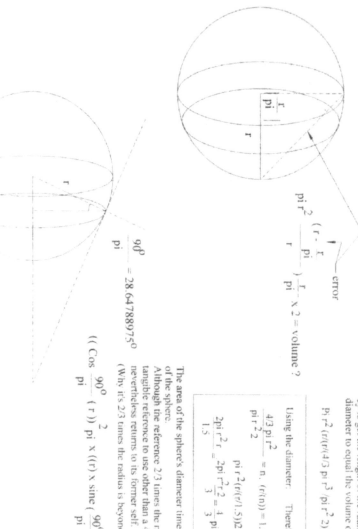

Using the diameter: Therefore, pi r 2(r/(r/(n))2 = 4/3 pi r 3

$$\frac{4/3 \text{ pi } r^2}{\text{pi } r^2 2} = n, \ (r/(n)) = 1.5 \text{ in all cases.}$$

pi r²(r/(r/1.5))2 = 4/3 pi r 3

$$\frac{2\text{pi } r^2 r}{1.5} = \frac{2\text{pi } r^2 r}{3} = \frac{4}{3} \text{ pi } r^3$$

The area of the sphere's diameter times 2/3 the radius times 2 = the volume of the sphere.
Although the reference 2/3 times the radius times 2 is used here to give a tangible reference to use other than a 4/3 and a cubed radius, the equation nevertheless returns to its former self.
(Why it's 2/3 times the radius is beyond me, just lucky, I guess.)

pi r 2 $\left(\dfrac{r-\dfrac{r}{pi}}{\dfrac{r}{pi}}\right)\dfrac{r}{pi}$ x 2 = volume ?

error

$$\frac{90^0}{pi} = 28.6478897^0$$

$$((\text{Cos } \frac{90^0}{pi} \ (r)) \ \text{pi} \ \text{x} \ ((r) \ \text{x sine} \ (\frac{90^0}{pi}))) \ \text{x2} = \text{volume}$$

2

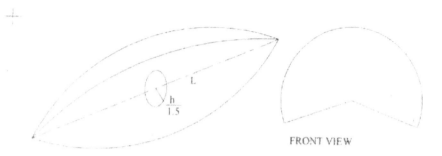

FRONT VIEW

$$\frac{n^\circ}{360^\circ} \times L \frac{h}{1.5} \times (\frac{h}{1.5}) \, 2 \, pi = \text{volume of the swept chord.}$$

Since this volume is the canopy its application of force in flight of WR applies to its center vertically.

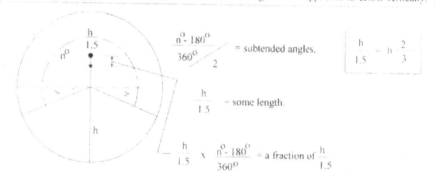

$$\frac{n^\circ - 180^\circ}{360^\circ} \frac{}{2} = \text{subtended angles.}$$

$$\frac{h}{1.5} = h \frac{2}{3}$$

$$\frac{h}{1.5} = \text{some length.}$$

$$\frac{h}{1.5} \times \frac{n^\circ - 180^\circ}{360^\circ} = \text{a fraction of } \frac{h}{1.5}$$

* = the center of the geometric shape for the canopy effected upon by WR with respect to the Em = 0.

All geometric shapes may be calibrated (indexed) to be balanced in simple terms using more simple of compound complex shapes.

If a complicated form arises, a more simple form may take its place to make the performance of calculations simpler.

These equations are hypothetical and there is not a proof with respect to the volume equation specifically referenced above.

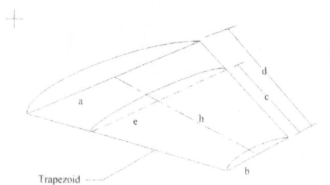

See equations page 13/65
Fig. 2-11.

Trapezoid

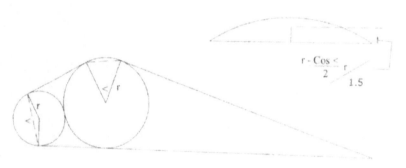

$$r - \frac{Cos <}{2} \quad r$$

$$1.5$$

Circles do not have to be tangent to three entities.

$<$ may be $360 / 2^n$, n may include a standard fraction (3-5/16, 3.3125). "r" may be a standard unit of measurement, e.g. 1.25".

$$\left[\frac{(pi\ r^2\ \frac{360^o}{2^n}}{360^o} \quad \frac{- Cos <}{2}\ r\ x\ \frac{Sine <}{2}\ r\ )\ x\ (\ r - \frac{Cos <}{2}\ r\ ) \quad x \quad \frac{Sine <}{2}\ r\ (2) = \text{area of the chord.} \right]$$

$$\frac{}{1.5}$$

$$\left[\ (\ a - b\ )\ c/d + b = e\ \right]$$

Multiply all the values of a base (a or b) by either e/a or e/b to get the values for chord e, depending on whether base a or base b is used for values to multiply to, determines whether e values are bigger or smaller than the chosen base.

Calculations for trigonometric geometric solids volumes may be made using base area, and center height, or center height area and total height. Trapezoidal geometric solids volumes require slightly more complicated equations.

A more effective wing shape.

WR

Formulas: 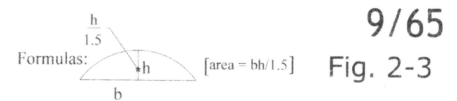 [area = bh/1.5] **Fig. 2-3**

The center of the chorded section is h/pi, determined from
the arc height to the chord.   1.5

This is the reasoning for the aerodynamic force
on the lift blade vanes and the operator's force at
the pedal:

Fig. 2-4

[A/Cos<1Cos<2=B, BCos<2Cos<1=A]

A-1024 pounds, B=2048 pounds when <=45 degrees.

The applied aerodynamic force at the lift blade
vanes is 2048 pounds. The applied force at the
impellor to drive the blade is an even number of
times greater, preferably 2x or 4x.

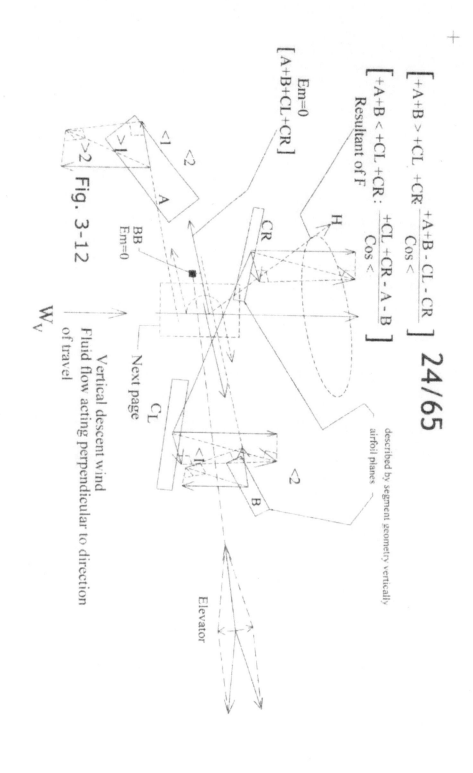

$$\left[\frac{+A+B > +CL +CR}{Cos <} : \frac{+A+B - CL - CR}{Cos <} \right]$$

$$\left[\frac{+A+B < +CL +CR}{Cos <} : \frac{+CL +CR - A - B}{Cos <} \right]$$

Resultant of F

Em=0
[A+B+CL+CR]

<1
<2

BB
Em=0

A

Next page

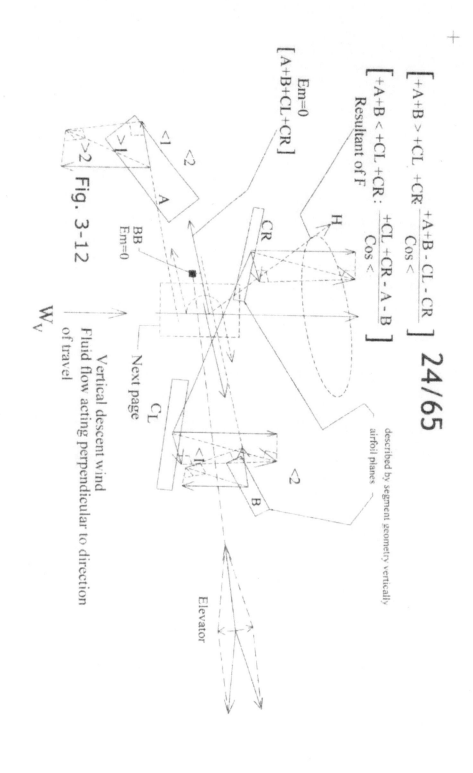

CR

H

described by segment geometry vertically
airfoil planes

CL

<1
<2

B

Fig. 3-12

W_v

Vertical descent wind
Fluid flow acting perpendicular to direction
of travel

Elevator

The requirement of lift is that it has to move under BB when vertical flow force begins and speed falls off to glide slope velocity. If lift is kept under BB' and there is no x value to WR at any time, then elevator controls vertical flow force sum of the moment =0 @ BB and horizontal flight has no vertical vector applied at BB; AA' is always aft of BB by Y2:Y1 or X2:X1. Reconfigurint the flimsing somponents any model aircraft, vessel, vehicle, or spacecraft may possess will reconfigure aA test bisector

repeatedly until Y1:Y2 or X1:X2 aligns BB congrunt to the lift vector xum of the moments =0 be moved **parallel** to the lift vector to align WR intersects BB'. There, done. Scale the airplane Mgm v design big enough to incorporate the required accessory components and design the mechanical workings internally and externally and the structure and whatever else you want and design another one. Good luck.

This configuration still requires elevator control nose down to get vertical air flow to align sum of the moments =0 with respect to WR by vertical times y - WRx = 0 when lift is subtracting all the vertical air flow force.

So therefore it becomes obvious that a body having the lesser volume leading the greater volume flies back'wards. The greater volume must be the leading in WR to have the equal volume bisector plane leading AA'. Engineering swept forward wing aircraft requires an addition of a fuselage to include leading volume greater than trailing volume.

The required analysis variables for balancing an airplane model:

AA' bellcrank static balancing
WR (displacement distance of one Mgm v)
vertical (displacement distance of one Mgm v)
1) aA test for vertical point
1) aA —
1) bB   — AA'
2) bB test for vertical point
graph
balance BB with cfm weights on resultant of WR and vertical from AA' at test bisector plane intersection.

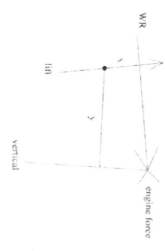

WR

lift

vertical

engine force

bB ——— @ AA'
aA

bB test bisector

N. Material gas blade vanes volume (Mgbv) 2 = "engine thrust force performance", and lift (which is part's volume divided by entire vessel volume (aircraft volume) times WR vertical times the resulting vertical vector magnitude of the airfoil plane angle of the part volume divided by the WR times air density) are included where engine thrust force performance is congruent to WR and lift intersects BB perpendicular to WR.

Precalculating subsegments volumes for prefabricated assembly should reduce calculating time. Standardization of prefabricated configurations should reduce calculations time. Dimensions of geometry is standardized. Parts come in fixed sizes and are interchangable.

Maintain that Subsegment (S) volume displacement unit quantity is fixed and WR is moving and not moving at the same time. (Conscript that in your mind?) (WR vertical) acts on S no matter what and S does not change anything. S does not act on WR. The displacement of WR is fixed in time and 2"a or whatever for S or whatever may be reevaluated to apply to first page specifications. If the displacement distance for WR = 1 Mgm v then any subsegment volume displacing the same distance displaces its proportion of subsegment swept volume of WR so every coefficient fold proportion multiple to every S is correct with respect to every S, not WR vert.

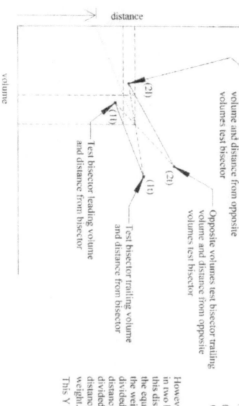

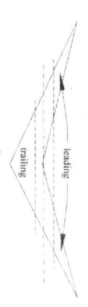

distance

volume

leading

trailing

nearly equal leading and trailing volumes
test bisector planes – – – – – –

Opposite volumes test bisector leading
volume and distance from opposite
volumes test bisector

Opposite volumes test bisector trailing
volume and distance from opposite
volumes test bisector

Test bisector trailing volume
and distance from bisector

Test bisector leading volume
and distance from bisector

(2l)

(1l)

(2t)

(1t)

When making a glider of such configuration AA' balances leading and trailing volumes almost equal and their centers distances are what are more different. When the graph is drawn there is no intersection of lines in criss-cross and the lines will not criss-cross, so another test bisector must be applied which is to make the leading and trailing volumes vary in size the opposite of the other test bisector volumes. This will make two lines on the graph which will create a criss-cross intersection, only there are now two perpendicular intersections of the criss-cross intersection for the two lines with their parallels to the intersection of the criss-cross perpendicular lines which apply with respect to Y1 and X1.

(2l)aA	second test bisector	[leading]
(1l)aA	first test bisector	
(1t)bB	first test bisector	
(2t)bB	second test bisector	[trailing]

However, it is seen that the dashed lines of perpendicular have intersection in two locations for which a line may be drawn to the Y- intercept. this distance may be applied to the distance from the bisector which describes the equal volumes leading and trailing, for which BB' will be located. Then the weight of the model x the density of the material the model is made of, divided by the conversion factor from cubic feet to cubic inches, times the distance of Y at the dotted line, divided by the counter weight mass in ounces divided by the conversion factor of ounces per pound to pounds, equals the distance from the bisector plane of equal volumes to the counter force mass weight.

This Y value is theoretical and is yet untried.

In the case of birds, their lift vector sustains their equilibrium around BB as x with respect to W.R changes with flexing in the air currents in flight and the sum of the moments equal zero or what ever the intended change of attitude rotation required, and the tail is a trim control whil the wings do the major establishment of equilibrium. The sweep angle of the birds' wings beat do maintain the birds' equilibrium by sum of the moments equal zero at BB or including rotation for attitude in flight and the use of the plural is intended to maintain that the birds' wings act independently, at all times for full spherical control in flight.

The distance perpendicular to the x axis from the criss-cross intersection to the line joining I1 and I1 (a A bB) is determined to be the distance from A A' to BB on the resultant. The line joining (I1 and it is the line through A A', except for the wing model glider. The only reference I am able to ascertain so far is the criss-cross intersection volume line perpendicular to the x axis extends vertically to intersect ((I1, (I1) and the distance from the criss-cross intersection to the line is (A A', BB'), I will build a model plane and see. The distance from the criss-cross to the (I1, (I1 line is the distance from A A' to the point where the leading and trailing volumes are equal.

I would prefer to think that Y2 - Y1 = the distance from A A' to BB on the resultant, unless it's shorter another way.

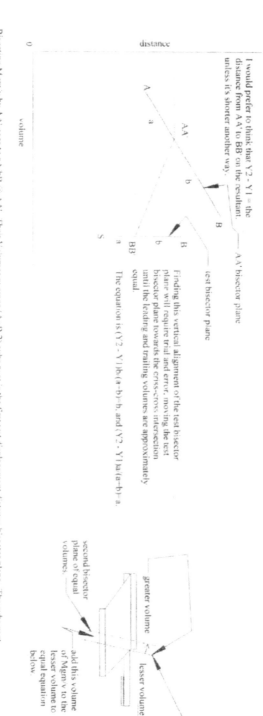

Finding this vertical alignment of the test bisector plane will require trial and error, moving the test bisector plane towards the criss-cross intersection until the leading and trailing volumes are approximately equal.

The equation is $(Y2 - Y1)b/(a-b)=b$, and $(Y2 - Y1)a/(a-b)=a$.

Bisecting Mgm v by A A' goes to a A bB @ A A'. Their Volumes average (A+B 2) cube root is the first test displacement distance bisector plane. The cube root distance should displace approximately $(A+B)/2$ cubic volume, add $(A+B)/2$ to the lesser volume and subtracting $(A-B)/2$ from the greater volume, although in this case the difference of b - a is .116 in. when cube root of $(A+B)/2 = .377$ in.

Applying a bisector line between A A' and the greater volume so that the volume between the two bisector planes equals leading (greater) volume - trailing (lesser) volume applies the half of the difference to the lesser volume and is more easily added than working a A bB test bisector, and the bisector plane distance from the A A' bisector plane is more easily discovered.

$$(\frac{A+B}{2})-B, \quad A+(\frac{A-B}{2})$$

(The swept volume of S = WR displacement distance

$$\frac{-2^n}{2^n} + n$$

S N_2 = lbs. at center of S. All aerodynamic subsegments also having airfoil plane angles

2   S

apply this displacement relationship.

$$\frac{Msm/v}{sum\ L} \times 2^a\ E_{sum\ S}\ N_2 = N_1\ Mgbvv\ ^{29}$$

This may be a tedious and time consuming problem.

L = vertical lift force of subsegment's S at airfoil plane angle developing from WR displaced with respect to coefficient fold proportion alignment limits of tolerances.

Mgbvv = material gas blade vanes volume (propeller blade vanes displacement volume having air density).

Es = sum S (summation of S (subsegments volumes individually,) having their respective a exponent per each.

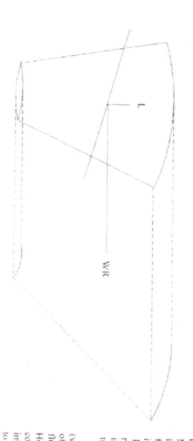

L

WR

Each 2^a Es applies with respect to each Es, and every 2^a S is multiplied by $\frac{Msm/v}{sum\ L}$. As with the glider, the only requirement of lift force does not require $\frac{Msm/v}{sum\ L}$ because only WR applies displacement distance on account of the aircraft isn't generating any power to sustain flight.

One unit vertical displacement volume is made equal to one Msm/v weight, which while the effect of slip stream air flow is (vertical/WR) x vertical congruent to WR with respect to wind flow force congruent to WR is equal to 1 Msm/v weight while its "momentum" (I use the term loosely here) is 1 Msm/v x (WR/vertical) for lack of reference. Say, the wind resistance is 1, the impact reaction force of the airplane model is approximately 28x Msm/v. Thus, the equations above are true.

(vertical/WR) x vertical congruent to WR is the coefficient of wind flow force congruent to WR with respect to wind flow force congruent to vertical simultaneously in equal time. However, if Mgm/v is accelerated to 1 Msm/v air flow force congruent to WR and vertical also equals 1 Msm/v, then the impact force of Msm/v at 1 Mgm/v air flow force congruent to WR is W.R/vertical x Msm/v weight.

Once the model plane balance is OK, then applying the aligned 2^p to ascertain a best flight performance will be allowed. Also, vertical displacement rate may not deliver 1 Msm/v impact force over vertical, so $\dfrac{WR}{vert}$ x vert. impact force = product of vertical impact force congruent to WR in equal time.

Let's assume that 1 vertical displacement of 1 Mgm/v distance = 1 Msm/v impact reaction force.

$\dfrac{WR}{vert}$ x $\dfrac{1 \text{ volume of S}}{S}$ x $\dfrac{2^n}{2^p + n}$ $S \, N_1 = S \, N_1 \, 2^n \, Mgbv.$ I'm not real sure how I arrived at this because I was falling asleep.

Vertical displacement in t time. S displaces WR in t time simultaneously as vertical displacement goes to zero, sum of L lift on glide slope increases to Msm/v weight force and sum of L(vertical) x (Msm/v divided by sum of L, vertical) = sum of L normal level flight.

$t \longrightarrow \dfrac{t}{Msm/v}$ = t displacing WR at normal level flight.

$\left(\dfrac{t}{\text{sum of L}_{vertical}} \right)$ Applied fold force of air flow to S is increased by   (

$\left(\dfrac{\text{swept volume of S over WR}}{S} \cdot \dfrac{2n_1}{2^{n_1}} + n \right) + \left(\dfrac{Msm/v \text{ weight}}{\text{sum of L}_{vertical}} \cdot \dfrac{2^{n_2}}{2^{n_2}} + n_2 \right)$
 on glide slope $ S \, N_1$

So, assuming that 1 vertical displacement of 1 Mgm/v applies 1 Msm/v force, then 1 WR displacement of 1 Mgm/v volume applies 1 Msm/v force, so S subsegments of airfoil plane angle apply S/Mgm/v x Mgm/v force over the swept volume of S/WR. Therefore the subsegments S applied load coefficient proportion are harmonious with WR and Vertical Msm/v applied load proportional swept volume displacements.

(S volume x .07651 lbs/cubic foot. 12 cubed cubic inches/cubic foot) x S/Mgm/v x Cosine of angle 1 x Cosine of angle 2 = 1 s_x.

Solve:

A A'.
WR distance = 1 Mgm/v
Vertical distance = 1 Mgm/v.

$\left(\dfrac{A - B}{2} \right) + B = A - \left(\dfrac{A - B}{2} \right)$ bisector plane distance addition of Mgm/v volume body geometry to BB from AA'.

$\dfrac{Msm/v}{2}$ x$_2^n$ sum sum $S \, N_2 = N_1 \, Mgbv \, 24$
sum L.

$\dfrac{Msm/v}{\text{sum L, vertical}}$ x $\dfrac{Msm/v}{\text{sum L, vertical}}$ = sum L normal level flight
sum L, vertical x

2

$$\left(\frac{\text{swept volume of S over WR or vertical}}{S} \cdot \frac{-2^{n_1}}{2^{n_1}} + n_2 \right) \left[-\left(\frac{Msm\ v}{\text{sum } L\ \text{vertical}} \cdot \frac{2^{n_2}}{2^{nc}} \right) + n_2 \right]$$

on glide slope

$= 2^P$ or 2^Q if the brackets are not included with respect to any sum of L, unless there is an L variable with respect to the geometry configuration of the aircraft model which there will be if more than rectangular and symmetrical aerodynamic work of airflow is performed.

(S volume $\times \dfrac{.07651}{12^3}$) $\times 2^P$ or 2^Q if vertical $\times$ Cosine of the angle 2 $\times$ Cosine of the angle 1 = L_{S_x}

Work bellcrank static balance beam until L_{S_x} sustain sum of the moments equal zero and parallel to WR and parallel to vertical vectors of sum of the moments equal zero intersect at the aerodynamic balance moment of the aircraft model mgm v.

cfm

c
A'
A

1:2

1:2

sum of L_S parallel to WR
bellcrank sum of the moments equal zero

sum of L_S parallel to vertical
bellcrank sum of the moments equal zero

In this case since the model aircraft is uniformly dense AA' is the volume of the model aircraft times .07651 12 cubed. Multiplied by c and divided by the counterforce mass, in this case .5 oz. 16 equals the distance from the location of BB' to the counterforce mass.

The displacements WR and vertical are still one Msm v effective applied force of one Mgm v swept volume displacement each for the parallels to WR and vertical. When describing vertical displacement distance when calculating 2^Q, 2^Q manifests itself in the ratio of swept volume displaced + S volume, not inches, as with 2^P, the ratio is not in inches, it's in number of times S goes into S + swept volume displacement. Since vertical swept volume displacement is accounted for 1 Msm v weight, then the same proportion of swept volume displacement congruent to WR equals 1 Msm v swept volume displacement in equal time. Any proportional fraction S thereof is equal to its proportional fraction. Therefore an error is found, and it is the trailing volume which is divided by S with respect to 2^P or 2^Q vertically.

The fact that vertical force x, y, / x = WR renders the fact that WR displaces vertical volume congruent to WR WR vertical times faster than vertical displaces vertical volume, despite the revelation in slipstream convergence. However I still have to prove that this static diagram is true.

It may seem possible that reversing the ratio of the centers distances of A and B with the opposite volumes will align the location for BB': a, b, A, B @ AA'. This is not so, because a is already less than b at AA': moving the bisector plane towards A center will never make b:a = B:A possible. The 2^P and 2^Q proportions multiply to Sx and when the sum of the moments = 0 aligns at intersection, then the 1:2 × 1:2 bisector plane aligns BB' where x with respect to y for the intersection vectors describes WR with respect to vertical distances ratio and a counterforce mass may be applied to align BB' congruent to 1:2 × 1:2 bisector plane intersects x and y where:

$$\frac{WR}{\text{vertical}} = x \text{ and } \frac{Msm\ v\ by\ BB'}{cfm} = \text{distance from BB' to cfm along Msm v by BB'}$$

aA /bB may not be congruent to WR. If not, rotate the AA' bisector plane around AA' until the line a.b is congruent to WR. Then, the 1/2 x 1/2 bisector plane may not be parallel to AA' bisector plane if vertical is at degrees angle to WR.

Never the less, AA' doesn't change and aA b = B is pretty much the same at any angle at AA'. The distance 1/2 x 1/2 bisector plane is made from AA' is taken along a. The line a.b is parallel to the sum of the moments equal zero aerodynamic equilibrium vector. It is easier to calculate (A - B)/2 between AA' bisector and 1/2 x 1/2 bisector Mgm v volume if the two bisector planes are parallel.

The applied vectors of Lx act around the moment of the intersection of the counter force mass (cfm) is prescribed, then the cfm is relocated initial location of the cfm prescribed from the aligning of the initial BB' cfm location is the fulcrum and the bellcrank force opposite Lx or the equal Msm v before the cfm location is aligned.

of the resultant and the 1/2 x 1/2 bisector plane until the initial location using the Lx vector sum of the moments equal zero acting around the to the resultant intersecting the 1/2 x 1/2 bisector plane. The initial cfm is the new location of the cfm. If Lx is lifting the aircraft, Lx must

Thus and so, if the static balancing of swept volumes displacement of Sx parallel to vertical, and Sx parallel to WR is not AA', then BB' does not always have to lead AA' but it does have to lead the intersection of Sx parallel to vertical and Sx parallel to WR and be congruent to the intersection of the resultant intersecting the 1/2 x 1/2 bisector plane. The leading volume has to be greater than the trailing volume.

This concept does not match equilibrium alignment.

If I do the ratio of the volumes backwards, B<A reverse A volume with B volume, the bisector plane may provide a solution because the 1/2 x 1/2 bisector plane is not providing the solution.

My other obvious oddball solution is (A-B)/2 -B at the AA' bisector add towards A the volume (A - B)/2 to B to the line (A - B)/2 + B or B + (A - B)/2, then find

resultant of Sx parallel to WR intersecting Sx parallel to vertical resultant 1 Mgm v swept volume displacement distances, align the resultant at the Sx intersection and where the resultant intersects B + (A - B)/2 bisector plane is BB'. AA' is somewhere else and the cfm must be applied to AA' if Mgm v is uniformly solid. to align AA' so BB' is on the intersection of B + (A - B)/2 intersects resultant from AA'.

Msm v exterior geometry should be continually redesigned with changes of calculating (A - B)/2 between AA' and (A - B)/2 to accomodate for any discrepancies in geometry which would change (A - B)/2 if Mgm v cfm geometry is exterior before reconfiguring for equilibrium moments.

intersection of parallels to WR and vertical of sum of the moments equal zero with respect to Sx lines swept volume displacement quotient.

The angle x/y = vertical / WR, of swept volume displacements distances of 1 Mgm v along parallels to WR and vertical.

9 degrees

WR

A

B

A'
B' R
a

A
A'
b

B

vertical

cfm

additional volume of Mgm v between AA' bisector and reverse ratio of volumes bisector /A - B)/2.

For aircraft models where B>A, the reversed bisector plane from the Sx sum of the moments equal zero intersection for WR parallel and Vertical parallel aligns BB' aft of the Sx's intersection:

Where lift is applied, lift is perpendicular to WR through BB'.

Sum of the moments equal zero.

initial Mgm'v volumes

Sx parallel to WR

ahead of aircraft

(A) (B)

aft of aircraft

bisector plane of reversed Mgm'v volumes

Sx vert.

cfm
vert
BB'
R
WR
B
A
AA'
d

It is now being tried where when A> B is trailing A. when the (A - B)/2 added to B bisector is trailing AA'; then the distance from AA' to the trailing bisector is reversed to front for the location of the intersection of the resultant of WR and vert. from the Sx sum the intersection moment. BB'. and the cfm is located with respect to AA' if the aircraft is uniformly dense (or BB if the aircraft is being assembled from parts).

The exterior weights' displacements moves the geometry of Mgm'v slightly. while moving the Sx sum of the moments equal zero intersection slightly. BB intersects Sx sum of the moments = 0 intersection moves cfm slightly. for this reason. BB is AA' in this case.

In order for a<b with respect to A>B, the leading volume at AA' must be >B. Therefore, the elevator should he a greater thickness. if the elevator is leading >B.

When correcting for A>B when B is > A be sure to be subtle because inexperience correcting for A>B to A>B going overboard on greater A will place the (A - B)/2 bisector plane far forward. Practice making less significant A>B alignments. Start by making A>B by using different dimension axes for A and B to align their geometry. Making a>b still requires more volume for A. Still, the location (A - B)/2 + B intersects Resultant for BB is yet to be an identity.

Construct model aircraft so A>B without changing its appearance much, or its equilibrium

Submarine buoyancy acts this way: The geometry's displacement of the subsegment's volume acted on by gravity creates a AB congruent to gravity. creating the foil plane which describes a A/b=B congruent to gravity. The vector displacement of force applied by the geometry volume of the subsegment is perpendicular to the foil plane at line a,b intersects foil plane.

The amount of the force of the force vector is equal to the volume of the geometry times the density of the displaced volume subtracted by the weight of the material makeup of the geometry. volume displacement.

Calculate: AA' Mgm'v geometric center and bisector plane at prescribed glide slope angle. Mgm'v volume in units. WR equal to one Mgm'v volume. Vertical equal to one Mgm'v volume, equal to parallel to glide slope vertical. Sx parallel to WR. Sx is subsegment volume displaced throughout one vertical or WR distance of its individual swept volume subsegment. Sx parallel to vertical, Sx is subsegment volume displaced parallel to vertical throughout the vertical displacement distance of one Mgm'v the same as WR, one Mgm'v volume displacement distance

(A - B)/2 bisector plane describing the reversed ratio of (B)>(A) = A>B Mgm'v geometry aligns these ratios.

Resultant at Sx intersection sum of the moments equal zero of all parallel Sx values parallel to their respective WR and vertical vectors magnitudes with respect to their perpendicular directrix, aligned to the values of WR and vertical as its components. Line (AA', cfm) intersection (A - B)/2 bisector plane intersection resultant)

Formula: AA' x (AA'. (A - B)/2 bisector plane intersection resultant @ BB'

weight counter force mass (cfm) equal units as AA'

= (A - B)/2 bisector plane intersection resultant, cfm distance).

A and B individual volumes.

Sx, having airfoil plane angle 2 x Cosine angle 1. reverse vector for cfm counter balance, apply distance to sum of vectors from initial location of cfm. apply formula: Sum of the moments of Sx equal to zero x distance from cfm to Sx/cfm weight equal units as sum of Sx = distance from initial location of cfm to new location of cfm from old location of cfm. Sx must be in equal units as cfm weight force. volume x density divided by conversion factor of cubic volume measurement.

Carefully whittling down any oversized A volume by subtracting the desired difference from A volume and dividing by two of the other dimensions you want to keep = the desired reduced volume of A. then finding Mgm'v volume - B = A and subtracting A - B and adding to B between AA' and test bisector plane until additional volume of Mgm'v subsegment = A - B. The volume. A - B. should be tried until the dimension distance between AA' and test bisector plane is small (reasonable). Now making A>B and keeping Mgm v geometry very similar throughout the design phase required making some of the volume thicker. However, it is discovered that the A:B ratio is very slight.

It seems that if A>B is dramatic, as opposed to if A>B is slight, then balancing BB' on (A - B) - B bisector plane intersects R-resultant requires Msm v airplane model, BB' being so far forward of Sx of WR intersect Sx of vertical, to fly very fast. Now if (A - B) - B bisector plane is engineered to be near AA' bisector plane, then Msm v airplane model does not require to fly fast as much, however, the former model plane should be made with only the least ctm applied so it is constructed as low density as possible. The farther away (A - B) - B bisector plane is from Sx of WR intersects Sx of Vertical the faster Msm v model airplane has to fly to fly, but (A - B) - B bisector plane has to be ahead of the AA' bisector plane to fly, frontwards, and BB' has to be at R-resultant intersects (A - B) - B bisector plane forward of Sx of WR intersects Sx of Vertical to fly right-side-up and frontwards, and R-resultant intersects (A - B) - B bisector plane has to be underside of WR to fly right-side-up.

With respect to one balsawood model plane 1 designed, I will probably have to increase the elevator volume until the A:B ratio is slight before I can eliminate the balance extension beam.

So I derived a model airplane where (A - B) - B bisector plane is described forward of AA' parallel bisector plane, and Sx of WR sum of the moments equal zero intersects Sx vertical sum of the moments equal zero is forward of (A - B) - B bisector plane. This presents a confusing and complicated solution: how to move (A - B) + B forward of Sx of WR sum of the moments equal zero intersects Sx vertical sum of the moments equal zero perhaps keeping AA' stationary, just move (A - B) + B bisector plane forward of Sx of WR sum of the moments equal zero intersects Sx vertical sum of the moments equal zero. My brain isn't a computer! I can't imagine or visualize how changes will effect the variables parameters. All that can be done is to vary the body geometry and calculate the parameters. There are no other variables.

So in fact it seems, that if A - (A - B) = B, then (A - B) + B = A and the Mgm v volume between AA' bisector plane and (A - B) bisector plane makes A - (A - B) = B, so aligning B geometry carefully (cell by cell) so B = A - (A - B) where the Mgm v volume between AA' bisector plane and (A - B) bisector plane makes A - (A - B) = B, equals (A - B) will require some computer programming, or meticulous repetitious equations resolving the same problems over and over until the closest proximity of decimal is humanly possible, which is usually two decimal places, while the accuracy of these models requires at least five decimal places of accuracy.

This relationship of A - (A - B) = B with respect to (A - B), makes the required geometry of B be a very slight variation from A while requiring Sx of WR sum of the moments equal zero intersects Sx vertical sum of the moments equal zero to intersect aft of the (A - B) bisector plane. Then again, the bisector plane of balance equilibrium may be A - (Mgm v volume between Sx vertical sum of the moments equal zero and (A - (Mgm v volume leading AA' = B bisector plane)) = B with respect to which A - (Mgm v volume from AA' bisector plane to bisector plane rendering A - differing volume being said = B makes the volume from AA' bisector plane to A - said volume = B bisector plane not equal to (A - B):

A vol. - B vol. = C vol.

: if B vol. + C vol. ≠ A vol.

and, A vol. - C vol. ≠ B vol.,

then the geometry of B molecules apply adjustment carefully ever so slightly to align A - C = B cell by cell if necessary hereditarily speaking; however,

A - B = C. But, A - C = B.

However, it appears obvious, that some model planes' A:B ratios may be very drastic and therefore aligning A with respect to B so their A:B ratio is very slight may be required, however some model planes drastic A to B ratio drasticness may only serve to balance the model plane instead of not, like when the wings are so far forward that when A - C = B aligns the distance between AA' and the plane of equilibrium is very great compared to when it is very slight, however, some planes may balance better in this case.

B vol.
C vol.
A vol.
AA' bisector plane
B vol.

The proportional ratio (A - B)/A = vertical/WR, or A/B = R/WR, however, this may be impossible it is only a theory, there may be a relationship between A, B, and C and WR and Vertical. Again, calculations will be repetitions and precision should be to at least five decimal places for scale models. Again, for living creatures cellular accuracy exceeds decimal places logic knowledge into the hereditarily adaptive strategy. What, exactly, is communicated to the genes regarding the equilibrium, certainly of any bird, is a mystery, because the bird doesn't know how to fly when it is born, it must learn how to fly. Nevermind. It seems to become sensible when C is created and there are two Bs, leading and trailing, when BB' is placed at C intersects R-resultant A with respect to B applies equilibrium where AA' applies Mgm/v equilibrium with respect to volume balance. A applies its proportionally greater volume to BB' where B + C = A (necessarily) in this respect in equilibrium by observation. The components are complimentary in equilibrium. are

But for the component for which the A:B ratio is drastic the other variables ratio limits of tolerances do not match, so the problem of ratio proportions is not obvious.

Calculate center of wing and fuselage rib areas
Calculate center rib dimensions and wings volumes
Calculate WR
Vertical
Sx of WR sum of the moments equal to zero
Sx of vertical sum of the moments equal to zero, both Sx's go to R-resultant
Find Mgm/v volume
Find AA'
Calculate B volume trailing
Calculate B volume leading (A volume less C volume, (A vol. - B vol.))
Locate BB' with respect to Sx of WR for all Sx of WR (having air foil plane angles) x .07651 lbs./cubic ft./12cubic inches per cubic ft. x Cosine of angle 2 x Cosine of angle 1 (sum of the Sx moments having lift equal zero force perpendicular to WR) x distance to BB'

BB'

= distance to new BB' location.

$$\frac{Msm/v}{Sx \text{ of WR} \times .07651/12 \text{ cubed} \times Cos >2 \times Cos >1} \times WR \times Sx \text{ of WR}$$ with respect to engine applied output inline blade vanes force N_1 Mgbvv 2^q, sum of the moments

equal zero @ BB': N_1 Mgbvv 2^q is congruent to WR, as Sx of WR goes to Msm/v-lift Sx vertical goes to zero.

Theorem: If the airplane can't fly with all the wheels on the ground the airplane can't fly. The airplane must be able to reach takeoff speed with all the tires contacting the ground and this provides the desired angle to make calculations for evaluating the minimum takeoff force applied in equal and opposite reaction to the force:

$$\left(\frac{Msm/v}{Sx \text{ of WR} \times Cos >2Cos>1}\right)(WR + \text{dead subsegment volume})(.06782\text{-}12^{-3}))$$

$$\frac{Sx \text{ of WR} \times Cos >2Cos>1 \quad \frac{-2^n}{2^n} \quad -n) N_1 Mgbvv = N_1 Mgbvv 2^q}{N_1 Mgbvv}$$

When making calculations for WR swept volume when calculating 2^F, only use WR displaced distance and one Mgm/v volume as before when calculating R-resultant because WR distance doesn't multiply but the product of unfolded time to displace WR volume applies 2^F to WR volume to equal the required force.

This latter formula applies lift and non-lifting subsegment body volume force compared to the displacement of the Sx volume compared to the density of air at 3000 ft. and the units are ((lbs/lbs)(cubic inches ÷ cubic inches)(lbs/cubic foot cubic foot)) lbs/cubic inches(cubic foot x cubic inches x lbs/cubic inch x cubic inches = lbs/cubic inch x cubic inches x fold multiple.

The steps for balancing a "glider", in this case a remote controlled glider, are not necessarily in this order:

align the components of the assembly so AA' is in the center of the design or the desired location for AA'.

define WR swept volume displacement and distance equal to one Mgm·v volume.

define vertical swept volume displacement and distance equal to one Mgm·v volume.

Calculate Sx of WR and vertical sums of the moments equal(s) to zero(s).

From Sx of WR intersects Sx of vertical, define R-resultant.

From AA' define B leading volume in cubic inches.

From AA' define B trailing volume in cubic inches.

Locate BB (actual model parts balance center before cfm) to the center of the design or the desired location with respect to Sx of WR intersects Sx of vertical, by locating the parts. This will relocate AA'. The opposite geometry of Mgm·v must be sized to align AA' to the desired loction in the model. This will move **BB**. The coordination of sizing the opposite Mgm·v geometry with respect to AA' alignment and the relocation of parts to align BB to its respective position with Sx of WR intersects Sx of vertical requires at least five decimal places of accuracy and several tries to align the points accurately.

Once BB has been located, then the cfm will align BB to B-leading intersects R-resultant.

Lift vector sum of the moments equal zero is counter rotated by relocation of 'BB' opposite of the lift vector.

The changing of the volume geometry of a less dense body volume component will move AA' more than BB. Increasing an outboard balsawood body volume component like an aileron size on one side of the airplane model) will move AA' more than BB' with respect to moving a servo motor inboard towards, the center of the airplane model) which will also move AA' but will move BB more than it will AA' comparatively with respect to the aileron density size change, and their alignment is hereditary with respect to the winged fowls. So aligning AA' and BB' on the centerline of desire for preferable heading "as the crow flies" these alignments should be carefully made.

Servos and batterys and wires and switches and control linkage and clevis pins and receivers and counter force masses and fuel tanks and engines etc all constitute the organs of a remote controlled model airplane. Still, the alignment of the components into balance may not be genetic but the accuracy of the hereditary content is analytically sound. The accuracy is subatomic to the infinitely small for decimal place accuracy, but there is no connection to creation, just analytical scientific observation of obvious facts.

So if you're ever intimidated by an entity to design something "real sleek", design something real sleek like a brick.

The rationale for lift devises Sx of WR Cos >2Cos >1 equal to the driving force of lift work being done on the glide slope. To sustain normal level flight then, it would seem that only

$$\left(\frac{Msm\cdot v}{Sx\ of\ WR\ Cos >2Cos >1} \right) = N_1 Mgbv v^{24} \quad if \left(\frac{Msm\cdot v}{Sx\ of\ WR\ Cos >2Cos >1} \right) Sx\ of\ WR\ Cos >2Cos >1 = Msm\cdot v, \text{ therefore.}$$

$$\left(\frac{Msm\cdot v}{Cos >1} \right) > [(WR - \text{non lifting volume} \times N_2)]$$

$$\left(\frac{\left(\frac{Msm\cdot v}{Cos >1} \right) > [(WR - \text{non lifting volume} \times N_2)]}{N_1 Mgbv v} \right) = 2^n \quad = n) \quad N_1 Mgbv v = N_1 Mgbv v^{24} \text{, but } \frac{WR}{Sx\ of\ WR\ Cos >2Cos >1\ Fm\ 0} - \frac{n}{1}$$

1 is variable with respect to n, 1 is a cubit, not a constant, 1 - y.

This represents the force effect of WR directed perpendicular to the airfoil plane as the required force of engine performance required to sustain normal level flight, not the equal to WR² as was previously described, however this is only a theory, since it is discovered that only Sx of WR Cos >2Cos >1 is required to sustain the glide slope.

Therefore the engine location for the side view of the thrust vector alignment placement with respect to BB' for the formula n-1 = WR Sx of WR Cos >2Cos >1 sum of the moments equal zero, aligns n with respect to BB' and lift sum of the moments equal to zero perpendicular to thrust at 1 from BB' with respect to n when WR Sx of WR Cos >2Cos >1 sum of the moments equal to zero = n.

It may otherwise be that WR/(Sx of WR Cos >2Cos >1 sum of the moments equal zero/Cos >1) = n, but that would make n smaller with respect to 1, but this latter equation is supposed to be corrected with respect to the thrust magnitude with respect to the lift magnitude so it is probably more right than the former equation.

Offspring can be called "alloys" of ancestral combinations of hereditary traits (alloys). Trying to keep all these combinations of formulas and rationale seperate made me accept that birds keep mixing them together to make more birds.

The displacement of WR, although it's direction is on the WR vector, is being directed perpendicular to the airfoil plane, so Sx of WR Cos >2Cos >1 is true so far but in theory Sx of WR Cos >1 = lift force, this will require some further consideration.

When the angle vertical makes with respect to WR is always the same. Msm/v must apply velocity along vertical and WR simultaneously to sustain the glide slope. So when Msm/v varies velocities vary, although the glide slope is the same. When WR is delivered to Sx the airfoil plane converts it to swept displacement parallel to the perpendicular to the airfoil plane so the vector magnitude appears to be smaller when actually it is equal to WR, because WR displaces Sx for a edge-on view and perpendicular to the airfoil plane is a face-on view of swept volume, but the swept volumes are equal, only the vector magnitudes are different lengths although they are the same magnitude, when they are applied to a computer and calculated for length they should be the same length. So let's say they are the same length. The aerodynamic balancing equilibrium of balsawood model gliders defines the glide slope as the swept volumes of WR and vertical at the angle attitude of the model plane on the glide slope using only the air density to define the glide slope R-resultant. Lift forces are individually calculated by only the air density volume of the subsegment volume having airfoil plane swept volume of WR at one material gas model vessel (Mgm/v) x Cos >2Cos >1

Now, ($\dfrac{Sx \ of \ WR \ Cos >2Cos >1 + all \ WR \ swept \ displacement \ subsegment \ body \ volume}{Cos >1}$) N_2 = applied engine force minimum

N_2 = .06782 lbs/cubic foot/12 cubed cubic inches/cubic foot, air density @ 3000 ft.

This formula is only using one swept volume displacement Mgm/v volume of air density, from which the formula ($\dfrac{Msm/v}{2 \quad N_2 \ Mgm/v}$ $- 2^n$ $/ 2^n$ $+ n)$ $\dfrac{}{N}$ $\dfrac{}{Mgm/v = N_2 \ Mgm/v}$ v_2^n was derived, finding the weight of the aircraft, Msm/v, a significant variable in the derivation for the formula. Now,

($\dfrac{Msm/v}{Sx \ of \ WR \ Cos >1}$ + WR)

$\dfrac{}{Cos >1}$ N_2 = applied engine performance directed along the sum of the moments equal zero of the lift vector resultant perpendicular to the airfoil plane = lift.

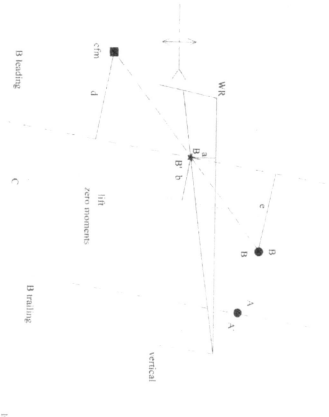

-e(BB) +bt(lift) + dx(cfm) = 0, solve for d.

This is now in glide slope mode.

When applying engine force performance the airplane resorts to level flight, therefore the glide slope parameters must be maintained in the event of running out of fuel and the necessity to have to glide safely.

Therefore a perpendicular from Sx of WR Cos >1

When the plane begins flying the sum of the moments with respect to zero goes away because there is no longer any vertical air flow force. There is lift, WR, and engine force all balanced around BB in three dimensions.

WR forces vectored to perpendicular at the air foil plane do not change their location of their applied moments so the sum of the moments equal zero does not change with respect to WR at the air foil planes. WR however, apparently applies in two directions at the same time at the points at the air foil planes where WR is perpendicular to the air foil planes and is also in the same direction it is coming from continually. Nevertheless, the sum of the moments equal zero with respect to WR at a and the sum of the moments equal zero with respect to lift zero moments at b and therefor the formula for normal level flight will be

-a(WR) +bt(lift) +-cx(engine force = WR + lift) = 0, solve for c.
                                   (not)

In this case the engine force vector is parallel to WR and the airplane model has to fly inclined to WR to sustain an air foil plane angle because very little of the model airplane has lift characteristics in its geometry configuration. The engine force vector is parallel to the air foil plane angle.

So I got it in my head start slow on the air foil plane angle with the geometric shapes, like maybe 1 degree of air foil plane angle, or what? With WR coming straight in in equal and opposite reaction force (e o rf) to an airfoil plane at 1 degree. Now calculate what the swept volume displacement is at 1 degree perpendicular to the airfoil plane. It should be displacing WR at 1 degree from perpendicular to WR, but if the airfoil plane is parallel to WR then inclines 1 degree the applied vector magnitude perpendicular to WR at 1 degree should be Sx of WR Cos >2 which holds true when Sx of WR x .07651 12 cubed = WR of Sx Cos >2
Cos >1 x .07651 12 cubed.

Top view apply r=a(WR) +-bx(yaw)+-cx(yaw engine) = 0
Solve for c. port or starboard yaw control.

So,
WR Cos >1 = Sx or summation of Sx individually.

F Sx + WR (Msm / ESx) = engine force.
Cos >1

Find Em = 0 with respect to Sx of WR and Sx of vertical.

Define R

Then B trailing, and B leading intersects R @ BB'

+/-a(BB) +/-(lift) +/-c(cfm) = 0, +/-x all go around BB'.

Then, three points contact on the ground, Em WR of Sx Cos >2Cos >1 = 0, there is no vertical air flow, the aircraft is on the ground, one Mgm/v volume displacement WR in the direction of travel, + WR swept volume including all non lifting subsegment volumes at air density for altitude. This total divided in to Msm/v = engine force performance, and therefore

$$\left(\frac{Msm\ v}{Em\ WR\ of\ Sx\ Cos >2Cos >1 + WR} \right) N_2 = \frac{n}{y}$$

n is the distance perpendicular to lift from BB', also described as b.

y is the dependent variable determined by equation of known values to space the engine force vector y distance from BB in equilibrium when the sum of the moments equal zero.

BB' is always located where B leading volume of Mgm/v to B trailing volume of Mgm/v, plane area intersects the Resultant of WR and vertical.

So you will notice, if there is no vertical air flow if the aircraft is sitting on the ground, there is no calculation for displaced volume perpendicular to the air foil plane except and the displacement is equal to WR of Sx one swept volume displacement and x Cos >2 equals the perpendicular vector magnitude to the air foil plane. Therefore the perpendicular to Sx air foil plane swept volume displacement can be calculated to equal WR of Sx x Cos >2. The geometry may then be calculated easily by ratio and proportion of the test volume and the calculated volume to get the actual displaced distance of the swept volume perpendicular to the air foil plane.

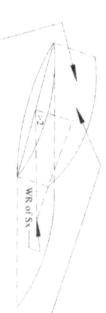

WR of Sx

Sx Cos >2 of WR = lift volume displacement parallel to Sx Cos >2.
(next page)

The calculation of this swept volume displacement is equal to WR of Sx x Cos >2.

Therefore, the appearance of lift is certainly perpendicular to the airfoil plane as is the swept volume displacement.

Therefore the previous rationale is rational.

Why WRN is divided into Msm v, is the equation for n/y such that WR of Sx is related to Msm/v so that the equation results in a dimensionless number only leaves me puzzling for any other rationale that WR of Sx could be associated with that would make that numerator dimensionless and not make y smaller with respect to n.

If this equation is done in cubic inches omit N_2, if in lbs. multiply N_2 through.

$$\pm a \underset{(n)}{Em\ WR\ of\ Sx\ Cos >2}) \pm b \underset{(lift)}{((Msm\ v)WR)} \pm y \left(\left(\frac{Msm\ v}{Em\ WR\ of\ Sx\ Cos >2Cos >1} \right) Em\ WR\ of\ Sx - WR\ \frac{Msm\ v}{Em\ WR\ of\ Sx\ Cos >2Cos >1} \right) = 0,\ \text{solve for } y$$

(drag)   (engine force performance)

+/- allows for location of vectors around BB'

19

As a taildragger aircraft applies all three contacts on the ground at normal speed at lift the Mgm-v WR vector xb torque may transfer from - to + or + to - as the aircraft levels out as it begins to fly. Mgm-v WRb follows the travel attitude and direction of flight of the airplane model. The torque will go from helping lift lift the plane on take of at normal speed at lift to stabilizing the increase of power at the engine force opposite BB or so to that effect, the sum of the moments never equal zero in flight due to the atmospheric surroundings only to complicate conditions of static equilibrium, reality however supposes that the sum of the moments equal zero at BB in all matters to be left for those interested in pursuing those endeavors. Such ideal conditions of flight are recorded here, conditions are fixed and unmoving yet remain alive.

System analysis:

Adding two plane's 3D parts geometry configuration and dividing by 2 will only produce slight variation in changes. Adding two different airplanes (ships, etc.) 3D parts geometries and dividing by 2 will produce a dramatic variation in changes. Saving desirable segments of certain models in the Selection Grid as an assembly will produce a new model. Saving desirable segments of certain models in the Selection Grid reproduces models with the desirable segments. The undesirable segments of the original assemble may be modified or remodifed until the original undesirable segment is popularized and then saved in the Selection Grid. This selecting dominant characteristics will eventually evolve to the superior popularity by appearances. Also, performance and aerodynamic "improvements" may be saved to the Selection Grid. Some artistic license may apply with respect to the sum of the parts and division by 2 since the computer may not allow an exact match for two different type craft. It may be necessary for the operator to control the match for the sake of essential components. The computer should give a preliminary sketch of the match before the operator allows it to be entered, although there is nothing that can change the preliminary sketch except reassigning a different component. In selecting two components a dominant component may be chosen, the recessive component will simply be made a percentage less than the dominant component. The recessive proportion to the dominant component may be selected.

So the swept volume displacement along the perpendicular to the airfoil plane will be $Sx \times Cos > 2$ of WR, with respect to the distance of WR swept volume equal to one Mgm-v volume displacement, corresponding to the WR distance after ratio and proportion alignment from test swept volume displacement alignment.

When Sx is flat to WR with respect to the air foil plane angle, no $Cos > 2$ is present so $Sx Cos > 2 = 0$. As the air foil plane is inclined to WR, then $Sx Cos > 2$ displaces $Sx Cos > 2 = $ some swept volume with respect to the trailing swept volume overall WR simultaneously in equal time.

$$+ \frac{Msm\,v}{EWR\ of\ Sx\ Cos > 2Cos > 1} \left(\frac{Msm\,v}{EWR\ of\ Sx\ Cos > 2Cos > 1} \right) \ - \ -b \left(\frac{Msm\,v}{EWR\ of\ Sx\ Cos > 2Cos > 1} \right) \times EWR\ of\ Sx\ Cos > 2Cos > 1\ N_2 + -\alpha WR \quad \frac{Msm\,v}{EWR\ of\ Sx\ Cos > 2Cos > 1}$$

$$+ \frac{Msm\,v}{EWR\ of\ Sx\ Cos > 2Cos > 1} \times EWR\ of\ Sx\ Cos > 2Cos > 1\ N_2) = 0,\ solve\ for\ c.$$

This equation should give the more accurate alignment for the location of the engine force vector with respect to the lift and the heading wind force.

c should not be a very large increment because drag and lift are not large increments from BB, which is where a and b increments are determined to be made from.

$c = \gamma,\ b = n,\ +\ _a(drag),\ +/-b(lift),\ +/-c(engine\ force\ performance).$

In my case, - (negative) is always clockwise and + (positive) is always counterclockwise, in a static torque diagram.

Also, $+ -a(WRN_y) +/- b(EWR\ of\ Sx\ Cos > 2Cos > 1) N_2 +/- \alpha WRN_2 + EWR\ of\ Sx\ Cos > 2Cos > 1) N_2 = 0$, which will derive the same answer for solve for c.

This equation does not include the quotient ratio of weight of the aircraft model with respect to the weight of the displacement of the subsegment body volume geometry having airfoil plane angle lifting the aircraft model.

dimension lines

Em = 0

WR of Sx Em = 0

cfm

B leading

WR of Sx Cos >2Cos >1

$-($EWR of Sx Cos >1$)b- -.BBd +-.cfmf =0$

glide slope configuration

Lift force is always perpendicular to WR and cfm is always opposite BB from BB' because BB is always at the intersection of R and cfm is always at the intersection of R and the bisector plane which describes

B leading volume

Vertical of Sx Em = 0

BB is not necessarily congruent to AA'.

AA' is the Mgm/v appearance geometry, and BB is the Msm/v material make-up of parts constituency weight.

Balsawood models are simply uniformly dense and BB is omitted where BB and AA' are congruent.

lift

B trailing

engine force performance centerline

taildragger take-off attitude

AA' bisector plane, B leading bisector plane, and vertical are all parallel.

Remember, if c is negative its torque is clockwise, and if its sign is positive its torque is counterclockwise.
Otherwise, you must remember to be consistent in whether your torque is clockwise or counterclockwise when it is negative or positive.

Glide slope configuration assumes the engine force equals zero. The changed attitude of Msm/v at three contacts on the ground at take-off speed changes the vectors with respect to their perpendicular distances from BB' and therefore c may have a value at take-off and for flying, since BB does not play in to the Em = 0 when calculating c, only lift and WR.

These are the variables to solve the powered airplane:
WR one Mgm/v volume swept volume displacement at N_2 air density, ... WR of Sx Em = 0
Vertical at one Mgm/v volume swept volume displacement parallel to vertical at N_2 air density. ..... Vertical of Sx Em = 0
AA': Mgm/v volume calculation
B trailing volume subsegment aft of AA'.
B leading volume subsegment equal to B trailing volume subsegment by bisector plane
BB aircraft weight of parts assembly
$+-a($lift$)_{bc}-b$WRN_2)$+-c$flift $-$ WR)$N_2 = 0$. Solve for c. a, b and c are all perpendicular to their respective coefficients.
R intersects B leading @ BB' for which the formula for the glide slope is $+-a($lift$) +-.b$BB $+-.c$cfm $= 0$, solve for c, a, b, and c are all parallel to WR
EWR of Sx Cos >2Cos >1 = lift.
The variables of length, a, b, and c do not coincide with the above drawing but b and c in this last equation can be replaced with d and f.
a, b, and c are used here merely for association. The coefficients, lift, WRN_2, BB, cfm, and lift - WR all have their own respective lever arm symbol.

WR swept distance needs to be described to align the swept volume equal to one Mgm/v before WR of Sx Lm = 0 may be applied. Also, when calculating take-off and there is no vertical in the balance, one Mgm/v swept volume is required to align WR distance before WR of Sx Cos >2Cos >1 may be calculated. Also, subsegments Sx balancing WR of Sx Lm = 0 at take-off attitude; three points contact on the ground, defining "WR" balance vector require the swept volume to equal one Mgm/v before calculations may be made with respect to WR of Sx Lm = 0

These, therefore, are the steps for balancing a powered model airplane:
Find:
AA' Mgm/v volume and center with all the parts compliment
Em = 0 WR of Sx @ 1lift attitude swept volume displacement of one Mgm/v
Em = 0 WR of Sx Cos >2Cos >1 @ one Mgm/v swept volume displacement
estimate +/-c(WR of Sx @ lift + lift) with respect to the alignment of the engine force vector
approximate c with respect to the estimated location for BB.
define WR of one Mgm/v volume displacement and distance for the glide slope attitude and define Em = 0 with respect to WR of Sx.
define vertical of one Mgm/v volume displacement and distance for the glide slope attitude and define Em = 0 with respect to vertical of Sx.
find B volume trailing at AA'.
define B volume leading equal to B trailing volume by bisector plane of c volume from AA' to BB' location.
using Em = 0 vertical displacement distance and Em = 0 WR displacement distance, find R-resultant and R intersects B-leading bisector plane @ BB'.
align c of +/-a(lift) +/-b(WR @ ground attitude) +/-c(engine force[lift + WR @ ground attitude]) = 0 exactly to side view.
align top view to c with respect to yaw vectors and Em = 0.
apply BBa cfm = b to locate cfm opposite BB. cfm-b-BB'-a-BB.

These steps will be employed in the balancing and alignment of my first remote control model airplane which I have to put the engine and fuel cell on and balance for it to fly.

All of these alignment variables apply with respect to the genetic matrix for individual airplane models.
If the fuel cell moves BB' from high to low when full then empty, balance the plane with an empty fuel cell. This way on take-off the operator will probably have to apply down elevator to pitch the nose down to prevent pitch up on take-off due to balancing the airplane with an empty fuel cell and having a high point BB' due to a full fuel cell being acted on by a large c value torque in the Em = 0 equation. The tail will probably drag at take-off speed with a full fuel cell because of the torque due to a large c value perpendicular to the engine force at BB'. The fuel cell, however, has to be centered on the B-leading bisector plane so BB' moves along the B-leading bisector plane as fuel is used up during flight and BB' moves from the high position to B-leading intersects R-resultant at fuel cell empty and the plane will glide on the glide slope according to the specified design parameters.

By matrix equations application, all the equations require their simultaneous equation where minor alignment of components to situate the values for the variables' coefficients changes the solution to the matrix so the application of changes aligns the Em = 0 with respect to all the variables values and coefficients simultaneously and the mortal can only calculate one equation at a time or only speculate on a simultaneous solution for all the equations, but a computer can be told to make minor adjustments to particular parts locations and have the required function to satisfy the simultaneous equations for each adjustment in the parts with respect to their position relocation developing a more satisfactory solution to all the equations simultaneously until the final position of the parts is the solution to all the equations.
This is a very time consuming problem for a mortal to do with a CAD program that simplifies the drawing, a pocket calculator, and some math background on paper.

Matrix of equilibrium

On model airplanes, the powerplant is sometimes naked mounted to the airframe, and the naked powerplant needs to be subjected to WR displacement distance and swept volume and vertical etc. and AA' Mgm/v volume, and B-leading volume with respect to B-trailing volume, and possibly +/-lift forces both up/down and yaw, and the engine usually applies yaw because of its exaggerated exhaust dilemma, so the engine force centerline has to be offset on the side of the exhaust pipe so the additional volume of the airframe will counterreact the exhaust body of the engine and so much arithmetic is about to result for the introduction of the powerplant and the accompanying components concluding the collection of parts for my first remote control model airplane. But I deliberately do not have any engine mount bolts or blind nuts to bolt the engine mount to the airframe because I have to balance the model airplane engine in place before it can be secured for sure and I want to be sure it's in the right place and I don't want any outside influence effecting my judgement when it comes to my patience putting together my matrix of equilibrium for balancing. This also introduces the element of urgency to refrain from using the welding equipment before assembling the power-driven system, so the axis may be aligned with respect to the matrix of equilibrium equations before any welding secures the axis of applied engine force to the airframe.

The balance centers for a scale model and a full size airplane of the same model will be different due to the actual geometry displacements configurations between the scale model and the full size airplane of the same model because scale models do not use scale representations of the same components as full size models, so their centers and bisector planes and all matrix equations variables values and solutions will be slightly different.

Design aircraft for most antagonizing appearances;
subordinate subsegments in to groups of individual components organized by sameness and similarity.
A A' Mgm/v volume and center
WR swept volume displacement = one (1) Mgm/v volume, WR of Sx Em = 0 balance
Vertical swept volume displacement = one (1) Mgm/v volume, Vertical of Sx Em = 0 balance
B-trailing
B-leading
R intersects B-leading @ BB'
Points contact "on the ground" airplane:
WR of Sx Cos >2Cos >1 (1 one Mgm/v volume displacement) individual distances (b) from BB' = individual lift forces.

$$BB = \text{aircraft weight and center before cfm.}$$
$$+/- a(\text{lift}) +/- b(BB) +/- c(\text{cfm}) = 0 \quad \text{glide slope}$$

$.07651$ lbs/3ft. = air density at sea level.

+/-at WR: one Mgm/v volume (@ $.07651/12^3$)+/-b(lift) +/-c(WR + Elift) = 0   Elift = the sum of all the lift values.   WR = one Mgm/v swept volume x $.07651/12^3$.
[+/-b(lift) applies b = b, c, d, e, f, etc. depending on how many individual lift characters there are in the aircraft model's design.]
[All WR and lift forces are characterized by (1) one Mgm/v volume swept volume displacement with respect to WR and WR Cos >2Cos >1 and x (times) $.07651/12^3$
except where the aircraft is engineered to fly at altitudes where air density is less than at sea level.]

Getting closer to finishing the first remote control model airplane design: Glide slope equilibrium alignment is first.
A A' = Mgm/v volume   WR Mgm/v volume   WR of Sx Em = 0
B-trailing   Vertical Mgm/v volume   Vertical of Sx Em = 0
B-leading   R intersects B-leading @ BB
  B-leading: align fuel cell centered on B-leading plane.
  B-leading: roughly align engine force axis, α(WR + Elift) @ "BB'"
BB'

bWR at wheels contact the ground angle attitude for b with respect to perpendicular to BB' for WR or Take off displacement distance alignment limits.
WR of Sx Cos >2Cos >1 for each lift force vector magnitude at perpendicular distance from BB' (a thru n)of Sx.
α(WR at contact angle to the ground + Elift) Elift = the summation of all the +/- lift forces.   WR + Elift = lift plus displacement volume of one Mgm/v at one atmospheric
air density for altitude divided by 12 cubed cubic inches per cubic foot.

Record the most possible summation of data quantity of as much of the airplane as possible to be used as one unit of information and use as few units of information as possible to evaluate the static analysis problems remaining.

It is discovered that the amount of work performed by subsegments having airfoil plane angle being multiplied by 2^P Cos >2Cos >1 is also multiplied by WR and by the subsegment volume divided by the Mgm/v volume divided by (/) 100, all multiplied by their perpendicular distances from BB' + - the torque of BB' from BB', and +/- the torque of the cfm, solving for the cfm lever arm. This applies the following; that the applied force of air flow to the WR vector is equal to one Msm/v while the impact force may be greater, which causes the aircraft to glide in air the WR air flow force acting on Mgm/v is equal to one Msm/v air flow force. Therefore the 2^P value is in effect, and the coefficient proportion multiplier to the torque is a percentage therefore it is divided by 100; e.g. torque lever arm length x coefficient force x ($2 \cdot$Msm/v / N_2 Mgm/v - $2^P \cdot 2^P$ +n)WRCos>2Cos>1 λ subsegment vol./Mgm/v / 100 = individual torque to BB' by one subsegment with airfoil plane angle.

WR 2P

$$+/- (2\ N_2\ Mgm\ v \cdot \frac{2^n}{Msm\ v} \cdot \frac{2^n}{} \cdot - n) \quad WR\ of\ Sx\ Cos >2Cos >1 \times \frac{subsegment\ volume}{Mgm\ v\ volume} \times \frac{.07651\ air\ density}{12^3\ cubic\ inches\ per\ cubic\ foot}$$

x lever arm length to BB' = torque at BB'

WR

Each one of these is a torque for each subsegment having airfoil plane angle applied to BB'. The +/- summation of the torque of BB and the +/- c(cfm) = 0, and solve for c, will give the desired result for the location for the cfm.

So the same impact force would be true of normal level flight = (equal to) one Msm/v of air flow equal and opposite reaction force (e.o.rf) for WR at nlf (normal level flight), while the impact force ("momentum") is a fold multiple of the Msm v inertia cubit.

$$E\ (2\ N_2\ Mgm\ v \cdot \frac{2^n}{Msm\ v} \cdot \frac{2^n}{} \cdot + n) \quad WR\ of\ Sx\ Cos >2\ Cos >1 \times \frac{subsegment\ volume}{Mgm\ v\ volume} \times \frac{.07651\ air\ density}{12^3\ cubic\ inches\ per\ cubic\ foot} = Msm\ v\ normal\ level\ flight$$

This value should be in pounds.

quotient

(Sx)

WR of Sx volume / Mgm v volume

Initially one swept Sx volume of WR distance was used for the fraction subsegment volume / Mgm v volume and it was discovered this might be in error, so Sx of the actual subsegment of the Mgm v body is now tried and the resulting location of the cfm should be even closer to BB'. However, it is discovered that the location of the cfm is even farther away but not wrong, in fact it turn out to be in the right place. The model glides perfectly, therefore the use of the formula proves true. The formula on the preceding page however, is in error because of the use of the denominator 100 which is now not used.

The introduction of the fraction subsegment volume / Mgm v volume seems out of place but the performance of the glider appears to make it feasible and I will continue using the formula as long as my airplanes perform adequately, using it.

24

When balancing the powered airplane model solving for the various parameters will result in five of the variables, requiring simultaneous resolution repeatedly until the c value is resloved for the engine force axis location:

WR = Mgm·v distance swept volume displacement on the glide slope attitude

Vertical = Mgm·v distance swept volume displacement on the glide slope attitude

and these are the five equations:

WR of Sx Em = 0

Vertical of Sx Em = 0

B-trailing

B-leading intersects R-resultant @ BB'

+/- a WR (Em = 0) take off attitude on the ground +/- "b" WR take off attitude, of Sx Cos >2Cos >1 +/- c (WR load + EWR take off, of Sx Cos >2Cos >1) =0, solve for c.

Solving for the swept volumes and the engine force axis and aligning BB' all together simultaneously requires reworking the equations over and over. If the equations were made elementary to a computer the solutions would occur more rapidly.

Make WR swept volume sketches for glide slope two then one forward and down subsegments at a time when calculating bellcrank static equilibrium working down from forward top to bottom leading to trailing. Do the front first and work aft and down. When solving for vertical, work from forward bottom first and work up and aft. The engine is a particular impediment in describing WR intersects vertical since the alignment of the engine force axis of power at c torque is required at BB'. The engine force axis of power may be translated = (WR + EWR of Sx Cos >2Cos >1). Still, BB' has to be defined to apply c, a, or b.

Recalling WR aflight applies the weight of the aircraft while the impact reaction force sustains the forward direction, so if the impact reaction force divided into the weight of the aircraft = sine of the glide slope with vertical, then even so the volume between B-trailing and B-leading (C) divided by B-(leading or trailing) = the same ratio as the weight divided by the impact rection force, and the glide slope may be appropriately, proportioned. This may be only a theory. The slower the plane goes the steeper the glide slope, and vice versa. But then these variables are not dependent because density is variable and BB' may be located anywhere arbitrarily, and glide slope may be engineered in.

When constructing the swept volume construction lines for the Sx subsegments it is possible to inadvertently apply the swept volume area center to the bellcrank line when calculating (a + b)(B(B +C) = b when defining the fulcrum of any ←——— bellcrank. When applying the calculated swept volume for any subsegment Sx at a non-specific WR displaced distance arbitrarily made, the applied volume is applied to the center of the Sx subsegment, not the center of the area of the swept volume.

In this case a:b = C:B since I use memory places identified by letters on my calculator, B and C are for calculated WR of Sx swept volumes, and (a + b) = the distance between the starting end point of the first of last calculated sum force and the next center of the Sx subsegment volume.

Repeat these equations simultaneously until the equation at the arrowhead aligns c with respect to Em = 0

(a + b)
a
b
c
B
B + C

Continue repeating this process of sum of forces and (a + b)(B(B + C) = b @ B + C until the final and last Sx subsegment is calculated and then work the vertical component Em = 0 until all the Sx vertical subsegment components are calculated. Then calculate Em = 0 B-leading, R-resultant intersects B-leading at BB', the ground contact WR of Sx Cos >2Cos >1 lift summation torques at BB' and WR torque at BB' and (Elifts - WR) @ c for engine axis of power for the Em = 0 for sustained flight. Good luck.

Remember, when calculating (a + b)(B(B + C) = b with respect to B + C, for calculating when calculating Vertical swept volume displacement distance using arbitrary distance initially, e.g. 1/2 inch or 1/4 inch etc. swept volume displacement arbitrary distance, the applied swept volume total of the subsegment's swept volume individually is applied at the center of the Sx subsegment not the swept volume area center. This facilitate the calculation of Vertical vector Em = 0 during the calculation of each Sx subsegment swept volume instead of calculating the swept volume in its entirety first and finding the unique distance of Vertical which is arbitrary distance sum volume divided by or into one Mgm·v volume whichever when multiplied to the arbitrary distance will produce the product of one Mgm·v volume at the true distance of swept volume displacement along the Vertical. Make sure the vectors of the swept volume quantity, magnitude for individual Sx subsegments apply to the center of the Sx sub-segment, not the swept volume area displacement.

To calculate the t/o (take-off) attitude (three points contact on the ground) of swept volume displacement the arbitrary displacement the trailing area of each Sx subsegment is calculated and the arbitrary displacement distance is calculated in, all the Sx subsegments are calculated and added to get the sum of the arbitrary swept volume. Then the known Mgm v volume is divided into or multiplied to the calculated arbitrary swept volume times the arbitrary swept volume displacement distance to equal Mgm v volume, and then all the Sx subsegments use the calculated swept volume distance of Mgm v volume with respect to Cos >2Cos >1 for each Sx subsegment times (x) their distance each from BB' (be sure to write each torque in a column on paper positive or negative for counterclockwise or clockwise respectively),and be sure when multiplying their arbitrary displacement distance to their Sx areas that you multiply, in Sine of the angle made at ground contact with respect to attitude to WR at normal speed at lift. Some geometry may appear difficult to evaluate for calculations but it is not difficult to "fudge" some calculations since by practicing model making it seems that minor calculation errors make negligible control errors in flight characteristics concerning the mechanics of the accuracy of the geometry calculations.

Right at starting calculating the arbitrary displacement swept volume in order to find a distance for a swept volume it is not within reason calculate any Cosine angles for any lift characteristics because every lift vector would have to be recalculated for the Mgm v displacement distance.

When evaluating the arbitrary swept volume for any Sx subsegment that's been subdivided in to sub-subsegments having airfoil plane angles, the description for the swept volume displacement for the arbitrary WR volume includes the entire volume of the subdivided subsegment subdivision with its perpendicular and includes any particular subsegment geometry included in the swept volume geometry:

All of these geometries calculations have depths towards or away from the sight area, for three dimensions to calculate volume.

The sum of the arbitrary swept volume is divided by or divided into Mgm v volume and multiplied to the arbitrary displacement distance for the swept volume displacement to equal one Mgm v volume.

These bodies have greater down force than these bodies have up force.

As the torque calculations, + - brSwept volume of WR of Sx Cos >2Cos >1, is being evaluated the summation of positive and negative values for the applied forces of Swept volume of WR of Sx Cos >2Cos >1 will define whether there will be the lift or not. If there is more force effecting up than there is down force summation then there is lift but the proportion of ratio of up to down forces must be a large number for lift to be at a slower velocity, at take off, so many of the down forces have to be minimized. My model has down forces that are presenting a problem of having greater force than there are up forces and I can only hope that when I have worked out the conclusion of the torques of "+/- lift" that I have somewhat more lift than I have down force. I still have some ways to go yet. I am more than halfway through calculating the +/- lift torques.

Be sure to mark which torques are forward of BB and which torques are aft of BB'. That way the counterclockwise torques forward of BB' are down forces and the clockwise torques forward of BB are up forces, and the clockwise torques aft of BB' are down forces and the counterclockwise torques aft of BB' are up forces.

Always remember to be consistent. If you make a mistake during dimensioning and you don't realize it until you have dimensioned several torques and it would probably be too much trouble to start over, just keep dimensioning the same mistake of dimensioning throughout the whole model and the problem should work out OK, as long as you are consistent with the same dimensioning error throughout the whole model and throughout the whole static problem, be consistent.

When solving the static equation for c with respect to the torque of the applied engine force axis with respect to BB', engine force torque may be clockwise or counterclockwise at Bs,' which will cause the + - sign to apply) to engine force torque at BB'. When moving c(engine force axis) to the opposite side of the equals sign the - or - will change sign to the opposite sign - or + respectively and if it becomes negative then you will have to multiply through the equation by -1 to change all the signs oppositely, respectively, so that c is positive.

Remember to locate BB' on the model and find BB by perhaps using a corner of a room or a window sill for a vertical reference and balance the model in more than one location to take a reference for where BB is. Look up where BB' is on the drawings and calculate the cfm weight BRa.b — cfm weight, where a and b are opposite lengths from BB' to BB and cfm respectively. BB----a----- BB----b----- cfm. The cfm may increase the weight of the model aircraft significantly. My model is increased in weight by 75 percent.

The following prospect are correct: the aircraft "house fly" defied flight. Once the characteristic alignments were in place the cfm was applied and the aircraft became too heavy to fly with the engine purchased for its flight, an .25 OS FP series MAX. The aircraft model requires a .95 engine. However I have endeavored another model airplane begining with the assundry characteristics;

AA' Mgm·v volume

B-trailing

B-leading

WR of Sx Em = 0, include engine and landing gear

Vertical of Sx Em = 0, include engine and landing gear

R intersects B-leading @ BB'

$+/-b(WR of Sx Cos >2Cos >1)$

Apply internal working components so BB is as close to BB as can be contrived. Use the balance (600gms. +/-1gm)

Apply cfm to align BB congruent to BB'

Align $+/-b(WR of Sx Cos >2Cos >1) +/-ccfm = 0$

Align $+/-aWR +/-b(WR of Sx Cos >2Cos >1) +/-c(WR + WR of Sx Cos >2Cos >1) = 0$ In both cases solve for c.

One swept volume of WR and one swept volume of Vertical apply simultaneously in equal time displacing equal force. so $+/- b (WR of Sx Cos >2Cos >1)$ is correct.

Model Airplane Design Plans

(actually fly)

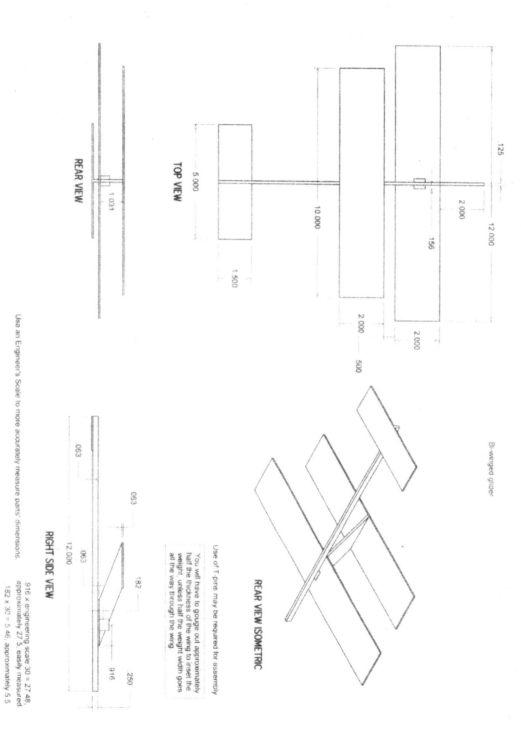

Bi-winged glider

TOP VIEW

125
12.000
2.000
156
5.000
10.000
1.500
2.000
2.000
500

REAR VIEW

1.031

REAR VIEW ISOMETRIC

Use of T-pins may be required for assembly

You will have to gouge out approximately half the thickness of the wing to inset the weight, unless half the weight width goes all the way through the wing.

RIGHT SIDE VIEW

063
063
12.000
063
182
916
250

Use an Engineer's Scale to more accurately measure parts' dimensions.

916 x engineering scale 30 = 27.48, approximately 27.5, easily measured

182 x 30 = 5.46, approximately 5.5

28

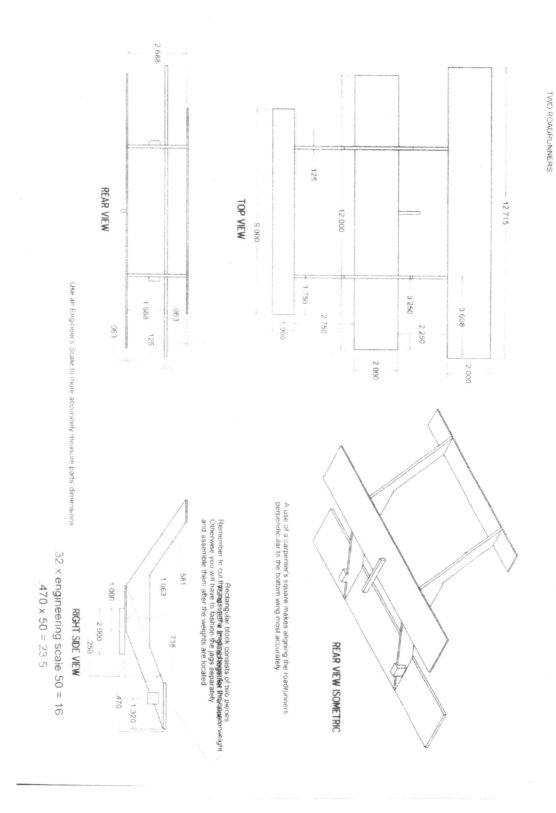

TWO ROADRUNNERS

12.715

3.608

2.000

12.000

3.250

2.250

125

1.750

2.750

1.000

2.000

TOP VIEW

2.668

9.000

063

1.688

125

063

REAR VIEW

Use an Engineer's Scale to more accurately measure parts' dimensions

A use of a carpenter's square makes aligning the roadrunners perpendicular to the bottom wing most accurately.

REAR VIEW ISOMETRIC

581

1.063

738

1.000

2.000

250

470

1.320

RIGHT SIDE VIEW

Rectangular block consists of two pieces
Remember to cut the pegs for the quarterweight together for the counterweight Otherwise you will have to fashion the jags separately and assemble them after the weights are located.

32 x engineering scale 50 = 16
.470 x 50 = 23.5

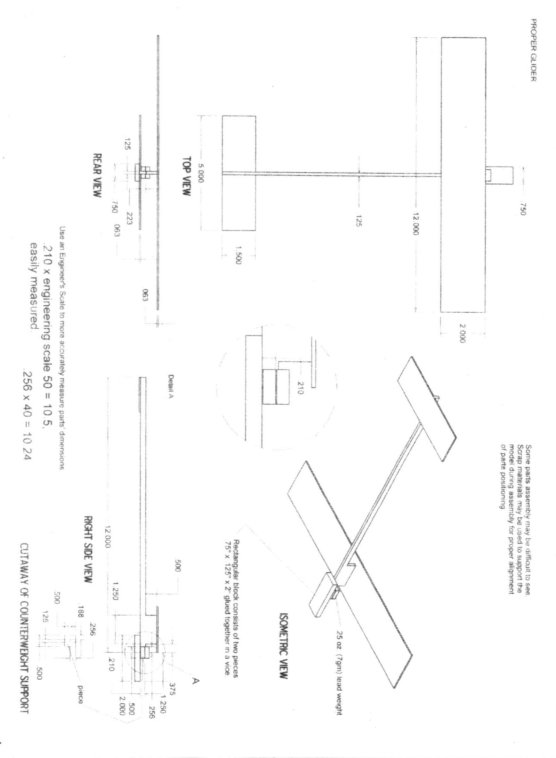

PROPER GLIDER

TOP VIEW

REAR VIEW

Detail A

ISOMETRIC VIEW

Rectangular block consists of two pieces
.75" x .125" x 2" glued together in a vice

25 oz (7gm) lead weight

RIGHT SIDE VIEW

CUTAWAY OF COUNTERWEIGHT SUPPORT

Some parts assembly may be difficult to see.
Scrap materials may be used to support the
model during assembly for proper alignment
of parts positioning.

Use an Engineer's Scale to more accurately measure parts' dimensions.
.210 x engineering scale 50 = 10.5,
easily measured.
.256 x 40 = 10.24

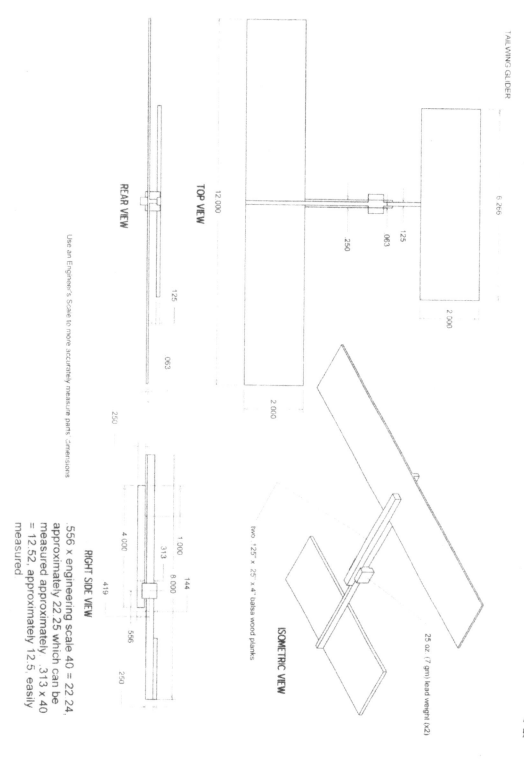

TAILWING GLIDER

TOP VIEW

REAR VIEW

12.000

6.266

2.000

2.000

.250

.125

.063

.125

.063

.250

Use an Engineer's Scale to more accurately measure parts' dimensions.

RIGHT SIDE VIEW

4.000

1.000

.313

8.000

.144

.419

.556

.250

.556 × engineering scale 40 = 22.24, approximately 22.25 which can be measured approximately. .313 × 40 = 12.52, approximately 12.5, easily measured.

ISOMETRIC VIEW

two .125" × .25" × 4" balsa wood planks

.25 oz. (7 gm) lead weight (x2)

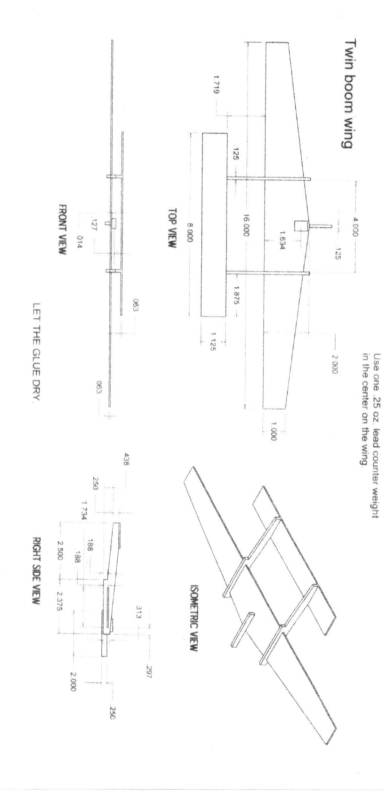

Twin boom wing

TOP VIEW

1.719
.125
4.000
16.000
.125
1.634
8.000
1.875
1.125
2.000
1.000

Use one .25 oz. lead counter weight in the center on the wing.

FRONT VIEW

.127
.014
.063
.063

LET THE GLUE DRY.

ISOMETRIC VIEW

RIGHT SIDE VIEW

.438
.250
1.734
.188
.188
2.500
2.375
.313
.297
.250
2.000

Multiply .634 x engineering scale 30 = 19.02. approximately = 19 and measure 30 + 19 to align the counterweight atop the wing.

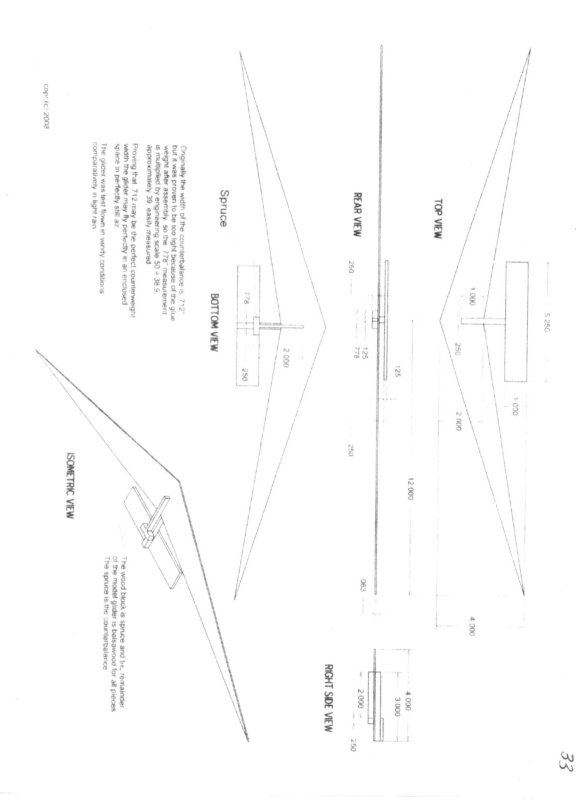

TOP VIEW

REAR VIEW

Spruce

BOTTOM VIEW

ISOMETRIC VIEW

RIGHT SIDE VIEW

Originally the width of the counterbalance is 712" but it was proven to be too light because of the glue weight after assembly, so the 778" measurement is multiplied by engineering scale 50 = 38.9, approximately 39, easily measured.

Proving that 712 may be the perfect counterweight width the glider may fly perfectly in an enclosed space in perfectly still air.

The glider was test flown in windy conditions comparatively in light rain.

The wood block is spruce and the remainder of the model glider is balsawood for all pieces. The spruce is the counterbalance.

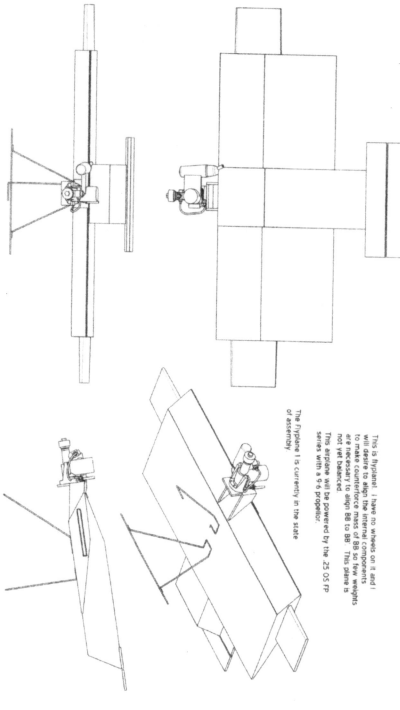

As can be made proven obvious, CAD software is deficient in the competency of fluid mechanics of inertia, this aircraft, spacecraft, vehicles, and vessels cannot be designed by CAD software due to their lack of competency in the understanding of engineering capability manifested by fluid mechanics of inertia.

I am Richard Chastain and I approve this remark.

This is flyplane! I have no wheels on it and I will desire to align the internal components to make counterforce mass of BB so few weights are necessary to align BB to BB. This plane is not yet balanced.

This airplane will be powered by the .25 05 FP series with a 9-6 propellor.

The Flyplane I is currently in the state of assembly.

Design Plan Formulation and

Mathematics

Fluid Mechanics of Inertia: the mechanics of flight, fly-by-the-seat-of-your-pants interplanetary space travel, the mechanics of floating boats and of sailboats (which operate in two mediums: air and water simultaneously), helicopter dynamics in static equilibrium, propetual motion, rocket science, submarines, virtually everything which a person can get in and operate, the solution to everything involving force and motion in the known universe.

Everything which lives and has breath involves the fluid mechanics of inertia equation:

$$NfD^{\left(\frac{\left(\frac{A}{BC}-D^e\right)}{D^{e+1}-D^e}+e\right)}=D^pNf$$

where D is any base, preferrably 2, e is an integer whole number, N is the displacement of a subject, and f is the subject's displacment's medium's density. A is Force in pounds. p is equal to the fold exponent.

$$\begin{array}{cccccc} D^0D^1D^2 & D^3 & D^4 & & D^e & D^{e+1} \\ 0\,1\,2\,4 & 8 & 16 & & \text{Some number} & \end{array}$$

Number Line

$\dfrac{A}{BC}$

BC may equal 1.

In the case of an airplane: A equals the airplanes' weight in "pounds", B is the airplane's displacement in cubic inches, C is the air density in pounds per cubic foot. A conversion factor of 12 cubed cubic inches per cubic foot is divided in to the denominator to get pounds per cubic inch. Then cubic inches cancels, pounds cancels and the A/BC fraction is a dimensionless number and can therefore be calculated for all of the powers. Since Nf equals pounds when the units cancel, the equation is in pounds since the exponent equation is dimensionless. The exponent equation will be described next.

p: this is the exponent of D. p is consistent when in base two it is proper to refer to it as "fold". In any other base it is not consistent to refer to p as fold. In base two when D = 2, when D raised to the p power x N x f = A an aiplane has an air flow wind load equal to its weight, Nf doubles p times to equal A.

$$Nf2^p=A$$

This is the mathematical foundation for Fluid Mechanics of Inertia. All rationale for every analytical math problem are solved using the formulas shown above, everything from blowing dust to interplanetary space travel. This formula may also be an integral genetic trait in the birds' heredity. This formula will be used to solve all manner of basic problems involving all problems of force and motion wherein a person can get in and operate a mechanism. Also the bird will be described to some detail, perhaps even fish, and the occasional insect. Feel free to take notes in the event you shall find something interesting to write down in your thoughts, perhaps a new invention or a question you would like answered and read on, perhaps there will be a solution deeper on in to the reading.

I apologize for the characteristics of the drawings in this device to describe the intention of these explanations with more detail.

Let's start with a boat floating on the water: the waves are in continuous motion, the boat is rocking with the waves. The facts are: the displacement of the hull at the "waterline", the rocking of the boat, the fact that the geometry of the water displaced by the hull continually changes as the boat rocks in the water, the vectors created by these variables, how to describe these vectors, where to locate a ships balance center, what the best geometric configuration of a ship design might be considering these variables, and any other considerations which may be consistent with engineering principles in ship design.

A ships hull at the waterline displaces a moving volume of water which may be bisected by a plane which opposite volumes have centers. Rotating this plane will locate the line joining the planes' opposite volumes' centers line perpendicular to the plane. As the water moves and the vessel moves in the water the planes moves, thus the line joining the centers moves and is always perpendicular to the plane. Now the vessel may be segmented to have many displacement volumes along the length of its hull. This will give many planes and many perpendiculars to the planes as the water is moving so much variation is water motion may be calculated. Since a ship is supported along the entire length of its hull by the water on both sides subjecting it to such a number of vectors is reasonable. As the water moves being displaced by the hull the perpendicular vectors move respectively as well as with the ship rocking in the water. Should the vectors magnitudes and directions be below the center of gravity and exceeding the torque required to oppose the topside of the vessel the vessel will capsize.

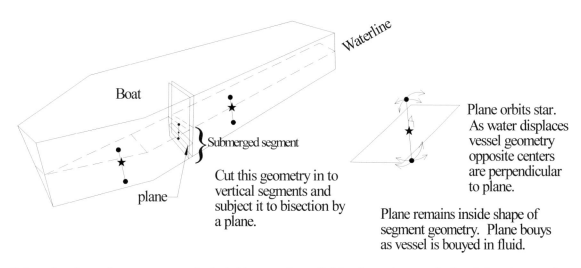

Boat

Waterline

} Submerged segment

Cut this geometry in to vertical segments and subject it to bisection by a plane.

plane

Plane orbits star. As water displaces vessel geometry opposite centers are perpendicular to plane.

Plane remains inside shape of segment geometry. Plane bouys as vessel is bouyed in fluid.

Submarines bouy the same way only their bisector plane divides the entire vessel when it is submerged. Aerodynamics plays in in motion in this case and will be explained later. Individual segments of the submarine may be subject to bisector planes individually having their own opposite centers line perpendicular to the plane. There is a left and right sides to the submarine. Constructing the geometry to apply the opposite centers lines perpendiculars to their planes to converge to a vertex above the submarine may provide some stability, or otherwise if the opposite centers lines perpendiculars are parallel there may be more stability. A submarine is less subject to waves, rocking and rolling in surface seas. Still, surface engineering is performed on submarines.

A sailboat is the most difficult craft to design of all craft. A sailboat operates in two mediums simultaneously, air and water. While air is moving in one direction water is moving in another direction or not moving. In the most difficult case it shall be discussed that air is moving in a direction and water is moving in another direction.

Moving air acts on the sailboat as it would on any displacement body, with the formula shown in the introduction as though the sailboat velocity vector with respect to the wind direction were subtracted from the wind direction and the sailboat were in an elastic freeze moment in time simultaneously with the wind. The entire displacement of the vessel above water is in equal and opposite reaction force to the simulaneous elastic freeze moment of the wind with respect to each individual parts' geometry of vessel configuration. Each different part exposed to the exterior has a different displacement. The wind applies a different force to each part. Depending on the parts' geometry its airfoil plane may apply lift whether it be up or down. At any rate, on a sailboat the windward view and the displacement effects the most force.

One way to calculate how much force is being applied is to test the parts to find out knowing how much wind is being used and how much force is resulting, then being able to calculate how much force is in total by knowing the wind speed knowing the sum of parts on the sailboat above water. This leaves a knowledge of a better sail, one which displaces more volume and therefore applies more force.

The parts under the water react the same way as before. Any current direction is and force applied is also directly related to the equation in the introduction, just use water instead of air where density of medium is.

Calculating the applied force of the sailboat displacement above water by using the wind speed and subtracting the vessel speed knowing how much force is applied at each increment of wind speed will give the applied bearing load of the vessel applied by wind difference, e.g. if 170 pounds of force is derived from 20 knot winds to the vessel and the vessel is moving at 12 knots then 170 /20 x 12 - 170 /20 x 8 = 34 pounds of wind force acting on the vessel which is delegated to the displacement of the vessel in the water and all of its applied forces in applied static elastic moment in freeze time simultaneous equal and opposite reaction forces. $\boxed{170 - 170 / 20 \times 12 = 68 \text{ pounds.}}$

Another way to discover the proportion of equal and opposite reaction force is to discover how fast a displacement is going when it has an perfectly elastic impact reaction force equal to its weight. This applies to all things in the universe to affect this equation. Remember, the body impacting is the displacement medium, not the displacing body. There is a charge for the impact reaction force of the displacing body. The applied force coefficient is the unit cubit which unit cubit is described throughout this text as a governing increment applying to everything in the known universe. A cubit is not the length of a man's forearm, it is a unit of one (1) of anything, e.g. one impact reaction force, one unit of displacement volume, one rocket nozzle volume, one rocket weight, one operator's weight, several operator's weights as one, etc.
Since things are done here in three dimensions most units cubit are cubic or weights, a cubit may be an altitude, or an air density, or a time increment.
Nf is a cubit divided in to A. Therefore the denominator is a cubit in a simple fraction. Having a complicated denominator like pounds per cubic foot x cubic feet divided by 12 cubed cubic inches per cubic foot   resulting in pounds as the cubit
inches
renders the denominator simple. Cubits are simple so don't get upset yet. These mathematics are simple. After 21 years of studying them and coming up with all the solutions it is as it is written in the Holy Bibile...the books that are left to be written are so simple that a child can write them.

Let's start with a bicycle: a bicycle and his rider has a displacement of air of approximately three cubic feet. Let's say the operator applies 65 pounds to the crank at the pedal. The bicycle and his rider thus moves a distance displacing one unit volume of approximately three cubic feet of air in a time fraction increment not recorded here. If that one cubic volume of air weighs .23 pounds then .23 pounds of force applied to move it would displace it in one unit of time. Now the pedal length is 7 inches and the chainring is 5 inches and the rear wheel diameter is 27 inches, so 65 pounds applies 6.741 pounds to the road. Therefore the bicycle and his rider do not displace any distance at the applied force and the windload is simultaneous. How do you calculate this problem? Here is the solution:

65 pounds x 7 inches / 5 inches /13.5 inches (half of 27 inches) = 6.741 pounds.

$$\frac{6.741 \text{ pounds}}{3^3 \text{ft.} \times .07651 \text{ lbs.}/3_{\text{ft.}}} - \frac{2^4}{2^4} + 4 = 4.836 \text{ fold}$$

$$2^{4.836} \times 3 \times .07651 = 6.741$$

Your calculator will not do this Problem.

$$\frac{6.741 \text{ pounds}}{3^3 \text{ft.} \times .07651 \text{ lbs.}/3_{\text{ft.}}} = 29.369 \text{ displacements.}$$

If we did know the displacement distance of ONE of the approximately three cubic feet, then the speed of cubic volume when its reaction force at impact is equal to its weight could be calculated. Newtons is not the answer, neither are Kg N m/s or Nm, or any of that miscellaneous arbitrary garbage.   The answer that I am seeking is in pounds and the distance is in inches and it is either secret or nobody knows.

]

Therefore the rationale for changing gears is present in order to accomodate the coefficient fold force for the varying windload. Operator plus wind and you must apply more force to the road and perhaps slow down to do so. Nevertheless, the increment of displacement can be variable isfone is wearing loose fitting clothing or has loose items on one's bicycle. The displacement distance of one unit volume will vary if the shape of it is variable and this will have an effect on the output performance of the rider's input by "shagging" the rider, so to speak. It is better to have items streaming longways to the wind rather than perpendicular to the wind. Objects which flail in the wind create a variable windload effect which will eventually become very tiresome since it pulls and releases at random force on the operator. Objects which flail in the wind should be tied down.

So a bicycle rider applying 6.741 pounds to the road applies $2^{4.836}$ displacements simultaneously to the bicycle and his rider's displacement. The approximately three cubic feet is folded 4.836 times at 6.741 pounds and the effect is simultaneous.

This is a simple problem. A more complex problem involves a meteor, which upon entering the atmosphere experiences so much folding of the air flow problem that the meteor burns up from air friction. The calculation for this problem also requires the creation of a variable.

Meteors volumes displace atmospheres of density at momentum of the meteors simultaneously, and the atmosphere at the momentum of the meteor reaction braking force applies to meteor momentum which braking force is a fold multiple of the meteor momentum, and the braking density is a fold multiple of the density of the meteor. Each fold multiple relationship is with respect to the air volume displaced by the meteor and its density.

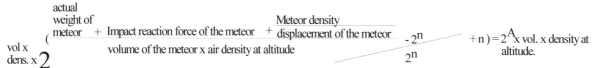

$$\text{vol x dens. x } 2^{\left(\frac{\frac{\text{actual weight of meteor}}{} + \text{Impact reaction force of the meteor} + \frac{\text{Meteor density}}{\text{displacement of the meteor}}}{\text{volume of the meteor x air density at altitude}} - \frac{2^n}{2^n} + n\right)} = 2^A \text{ x vol. x density at altitude.}$$

Also for every fold of 2^A with respect the volume of the meteor x air density at altitude I would assume the temperature of the volume of air being displaced will double 2^A times. until 2^A x volume x density at altitude equals the melting temperature of the material the meteor is made of. If the impact reaction force of the meteor exceeds this moment so much the better. If its impact reaction force is below this moment the meteor may become a meteorite and strike the ground.
So you see so far the simplicity of the use of the formula. Applying force and volume and density to the equation will result in a variety of solutions.

This equation also applies to interplanetary space travel in a fly-by-the-seat-of-your-pants fashion:

Introduce fly-by-the-seat-of-your-pants interplanetary space flight here.

The determination to find an inventive step from the formula for fluid mechanics of inertia to discern the generic appearance of a bird into the ability to determine the differences between the sexes is yet inoperative. In the Holy Bible our Father In Heaven took a "rib" from Adam, the generic prototype of all mankind, and made Eve, Eve was all inclusive of human details, containing all the variables and all their values and all the required solutions for complete and successful operation of a fully functioning human being of the appropriate sex, as was Adam successfully opposed by sex to the woman. It may be determined that a successful mechanism that will input data in to a male will not function for a female due to their mind's functional performance variations of soma clusters and their rational solutions separately. Males solve problems one way and females solve the same problems another way. Males do not bear the child. Females and males are not necessarily compatible in the solution of simultaneous equations. Males and Females function simultaneously and would require continuous monitoring and since they are autonomous and operate continually the variations in their equations are unique although they are similar in constituency make-up of materials.

The input for training is the same although subtle variations may make up the difference in the compatibility for understanding between males and females.

Further thought determines that it may be more feasible to discern the senses to be more applicable to finding values for variables to include. Still, an inventive step is required to find a soma cluster of variables to interconnect in to an equation with a solution to send to the chromosomes. I saw a vision yesterday evening of what may be a brain language on a monitor: it is horizontal and vertical lines of varying width with solid round ends of varying small sizes at the line's ends, and having various spacing.

These elements are flashed on the monitor in whatever variety is determined to be interpreted have meaning.

These could be soma clusters, their links could be different colors and hues, the solid circles could be different colors and hues, either side could be different colors and hues, etc.

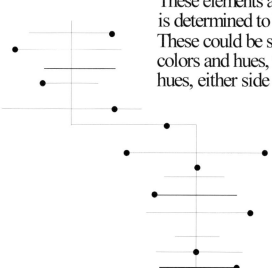

The location of each element could vary by so slight a difference that it could convey a different meaning.

Any variation however small could convey a different meaning.

Discover counting number, the alphabet, individual words, make sentences, do calculations, find the basic functions, start simple.

How does a soma or soma cluster take electro-chemical signals from the senses and interpret their meaning into something that makes sense to me?

How does a soma or soma cluster interpret electro-chemical signals at all?

What part of the formula sends the interpreted cognizable information to the part of my brain that understands the solutions if the solutions to the formula(s) go to the chromosomes as well?

How am I capable of comprehending what my brain interprets as chemical signals?
Where is this "I" in my brain?

48

Calculations for 140 mph hydraulic bicycle:

→ 7.14221 + 1.80856 +
4.76148 = 23.71225 in³
×8 = $\frac{209.69799 \text{ in}^3}{1.33658 \text{ in}^2}$
= $\frac{2 \text{ Rev/crank 6 R}}{4 \text{ PTUs.}}$

final impeller

.3125   .77773

3.018873 in + 2.01258

3.79663 in

5.80918 in/Rev

= 10.3592 in deep
$\frac{6.3731 \text{ in}}{.025}$ calculate slip
(12 in impeller)

primary
impeller

$\frac{73.76871}{73.76871 - 38.2496}$ ×
73.76871 = 72.653731
= 6.89441 in deep

$\frac{81.68741 \text{ in}}{12} = 6.80678 \text{ ft/Rev}$

$\frac{10 \text{ in} +}{31.89433}$
$\frac{1.85936}{1.962332 \text{ in}}$ (crank 6 R)

41.89438 in
$41.89438 - 3.82496$
$\uparrow 1.89438$ = 30.16596 Rev/sec

$\frac{140 \text{ mph} \times 44 \text{ ft/sec}}{30 \text{ mi/hr}} = 205.3 \text{ ft/sec} = \frac{205.3 \text{ ft/sec}}{6.80678 \text{ ft/Rev}}$ = 30.16596 Rev/sec

5.80918 in/Rev × 30.16596 Rev/sec = 175.23962 in/sec centers distance
1.96454

2.14859

.375 lb
$\frac{N}{\pi}$ = 1.90986 in dia.

1.33658 in²

$\frac{175.23962 \text{ in/sec}}{1.33658 \text{ in}^3/\text{Rev}}$ = 175.23962 in/sec
$\frac{2 \text{ PTUs}}{8 \text{ Revs}}$

6.19438 in deep (initial)

Calculate slip
5.22165 in deep
5.38768 in deep
.49505 in

$\frac{1.96966}{2}$ = .95493 in 1.5674   2.062453 in
46.95592 50.95093

$\frac{5 \times 48}{\pi}$ = 3.8197 in dia

$\frac{.5 \times 36}{\pi}$ = 5.72958 in
$\frac{2.86479}{.95493}$ = 3

$\frac{41.3803 \text{ lb}}{3:1}$ =
2:1 or greater @ the crank
6.89672 lb. @ the crank.

Geomagic.ISS.americas@
3DSystems.com

whole depth of tooth

ID O.D.

whole depth of tooth

$D_P \cdot .6866 =$ whole depth of tooth

$\dfrac{D_P N}{\pi} = D$

$\dfrac{(N+2) P_P}{\pi} = O.D.$

$\quad D_P$

N = tooth count

$N(a)$ = on 2 gear or the other

D = basic diameter

$\dfrac{(N+2) D_P}{\frac{\pi}{2}} = I.D.$

$(D_P \cdot .6866)2$

$\dfrac{(N_1 + N_2) \cdot P_P}{6.2832} = $ Centers Distance

.865 2.26796

whole depth of tooth

$\dfrac{D_P \cdot N}{\pi} = D$

$\dfrac{D_P}{.395} \quad .395 \cdot .6666 = $ whole depth of tooth

$\dfrac{(N+2).375}{\pi} = O.D.$

17 = 2.26796

whole depth of tooth

.8165 (1.90966) 2.14659

14.000
5.517

17T
ID 2.75305

.25748

$\dfrac{(N+2).375}{\pi} = O.D. = $

17 = 2.26796

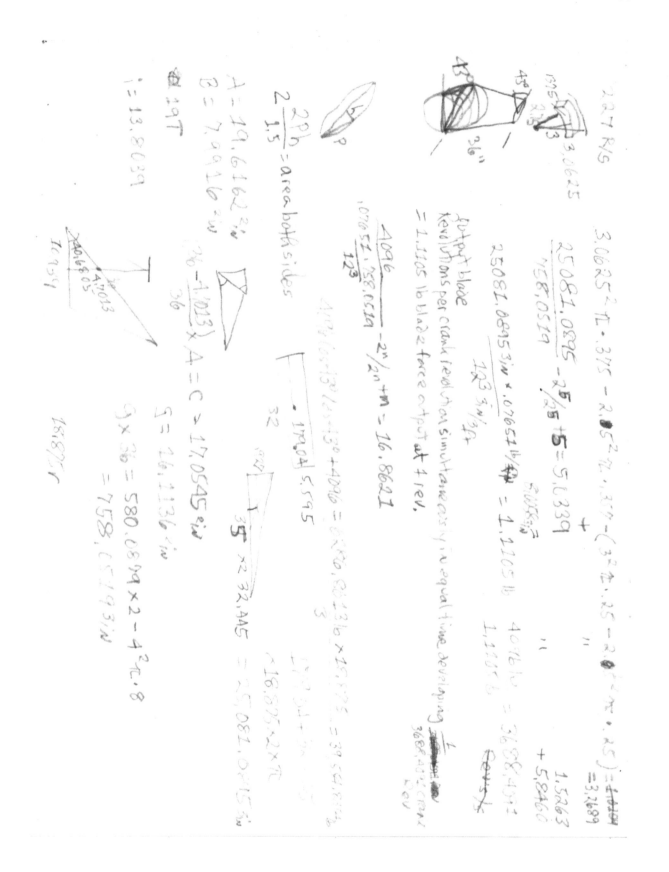

51

$$\frac{39554.8304 \, lb}{113 \, in^3/lb}$$

$$59.7 \, \tfrac{lb}{in^4}$$

$$\frac{8.6394 \, in^3}{113 \, in^3/lb}$$

$$8.0058 \cdot 16.8621 \cdot 19.5847 = \frac{2643.8438 \, in}{8.6394 \, in^3} = 12.7501 \, in \, deflection$$

$$8.0058 \cdot 1.5263 \, in \quad -2^n/2_n + n = 19.5847$$

$$8.6394 \, in^3 \quad -2 - n - 2 - 8.6394 \, in \times 24 \, Rev$$

$$\frac{8.6394 \, in^3}{24 \, Rev} = 4.2503$$

$$120 \, lb$$

$$27.8452 \, ft \cdot .375 - 27.5^2 \pi .25$$

$$27.8452 \, ft \cdot .375 - (27.75^2 \pi \cdot .25 - 27.5^2 \pi \cdot .25) = 9.5153 \, in$$
$$= 31.319 \, in$$
$$\times 2 = 42.0602 \, in$$

$$\frac{120 \, lb}{53.9 \cdot 9.5153} \quad -2^n/2_n + n = 8.5793$$
$$\frac{12^3}{}$$

$$42.0602 \times 24 \times 8.5793 = \frac{8660.3633 \, in}{(8.6394 \times 3 \times 6)} = 27.8452 \, in \, depth$$
$$\frac{}{2 \, pairs}$$

Calculate slip

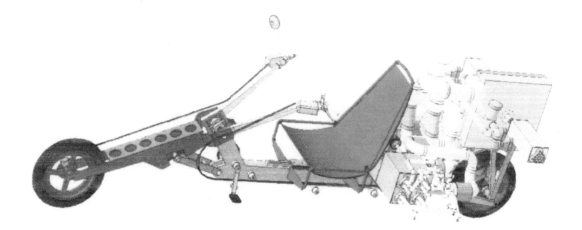

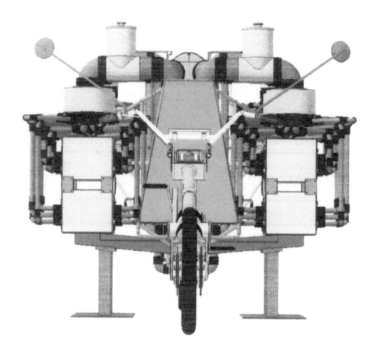

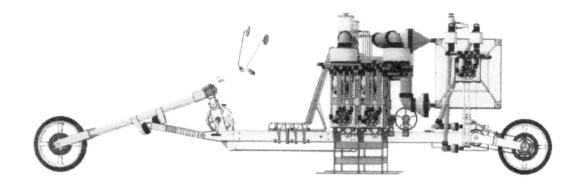

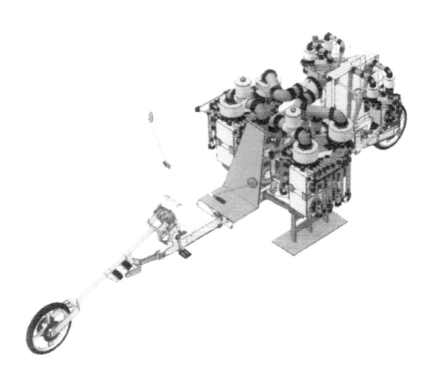

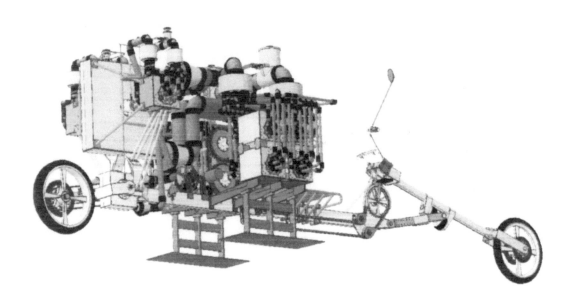

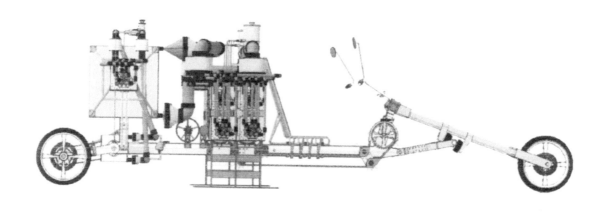

+

$$\left[\frac{-gL6 + gL5 + (g + 2e)L4}{e} = X\right]$$

$$\left[\frac{eL4 - eL5 - (e + 2g)L6}{g} = Z\right]$$

$$\left[L5 = L2_a + L2_b + L3 + L1\right]$$

$$[c > f, \; L3 < L1]$$

X, Z = left, right
tires road
contact force
Y = front tire road
contact force

$$\left[L2_a = L2_b\right]$$

$$\left[L3\frac{a}{b} + L1 - L2a\frac{d}{b} = L5\right]$$

$$\left[\frac{-L1f}{b-d} + L3\frac{c}{b-d} = L2_b\right]$$

$$\left[Y = \frac{fL1 + L3c + L2_b(b-d) + L2_a}{b}\right]$$

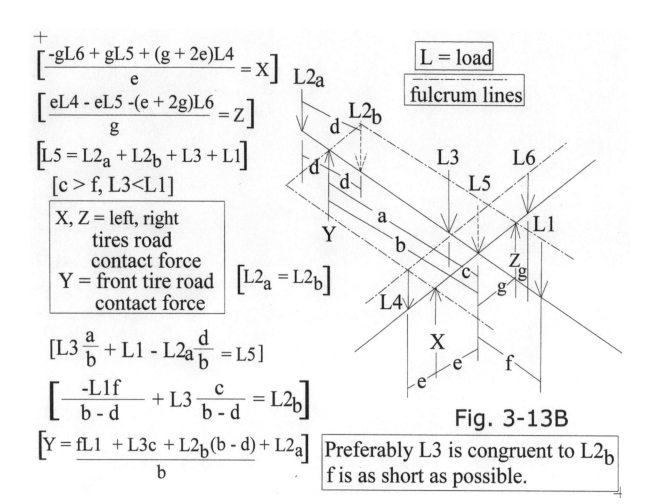

L = load

fulcrum lines

Fig. 3-13B

Preferably L3 is congruent to $L2_b$
f is as short as possible.

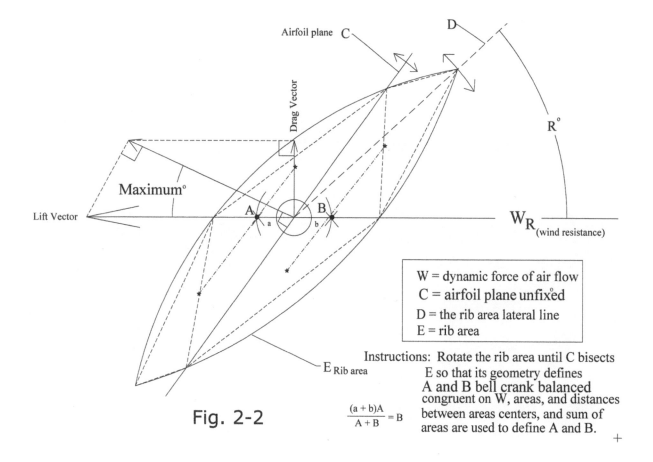

Fig. 2-2

$$\frac{(a + b)A}{A + B} = B$$

W = dynamic force of air flow
C = airfoil plane unfix̊ed
D = the rib area lateral line
E = rib area

Instructions: Rotate the rib area until C bisects
E so that its geometry defines
A and B bell crank balanced
congruent on W, areas, and distances
between areas centers, and sum of
areas are used to define A and B.

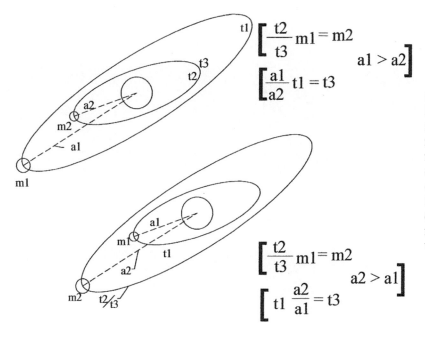

$$\left[\frac{t2}{t3} m1 = m2 \quad a1 > a2\right]$$

$$\left[\frac{a1}{a2} t1 = t3\right]$$

$$\left[\frac{t2}{t3} m1 = m2 \quad a2 > a1\right]$$

$$\left[t1 \frac{a2}{a1} = t3\right]$$

Calculations for celestial bodies
are in densities, and velocities
ratios multiplied to altitude
ratio in equation derives the
relationship for the change of
densities with respect to the
difference of velocity resulting
in this original equation: less
dense bodies orbit faster than
more dense bodies and vice
versa.

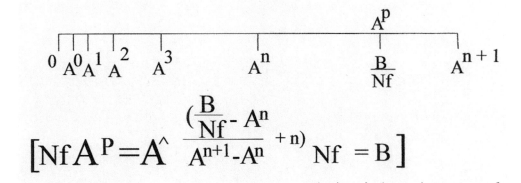

$$\left[NfA^P = A^{\wedge} \frac{(\frac{B}{Nf} - A^n}{A^{n+1} - A^n} + n)} \, Nf = B \right]$$

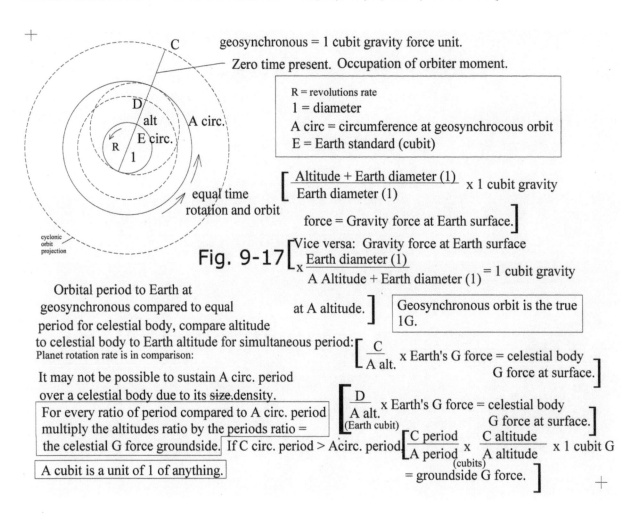

geosynchronous = 1 cubit gravity force unit.

Zero time present. Occupation of orbiter moment.

R = revolutions rate
1 = diameter
A circ = circumference at geosynchrocous orbit
E = Earth standard (cubit)

equal time rotation and orbit

$$\left[\frac{Altitude + Earth\ diameter\ (1)}{Earth\ diameter\ (1)} \text{ x 1 cubit gravity} \right.$$

$$\left. force = Gravity\ force\ at\ Earth\ surface. \right]$$

Fig. 9-17

$$\left[Vice\ versa:\ Gravity\ force\ at\ Earth\ surface \right.$$
$$x \frac{Earth\ diameter\ (1)}{A\ Altitude + Earth\ diameter\ (1)} = 1\ cubit\ gravity$$

cyclonic orbit projection

at A altitude.]

Geosynchronous orbit is the true 1G.

Orbital period to Earth at geosynchronous compared to equal period for celestial body, compare altitude to celestial body to Earth altitude for simultaneous period:
Planet rotation rate is in comparison:

$$\left[\frac{C}{A\ alt.} \text{ x Earth's G force = celestial body} \right.$$
$$\left. G\ force\ at\ surface. \right]$$

It may not be possible to sustain A circ. period over a celestial body due to its size.density.

$$\left[\frac{D}{A\ alt.} \text{ x Earth's G force = celestial body} \right.$$
(Earth cubit)
$$\left. G\ force\ at\ surface. \right]$$

For every ratio of period compared to A circ. period multiply the altitudes ratio by the periods ratio = the celestial G force groundside. If C circ. period > Acirc. period.

$$\left[\frac{C\ period}{A\ period} \text{ x } \frac{C\ altitude}{A\ altitude} \text{ x 1 cubit G} \right.$$
(cubits)
$$\left. = groundside\ G\ force. \right]$$

A cubit is a unit of 1 of anything.

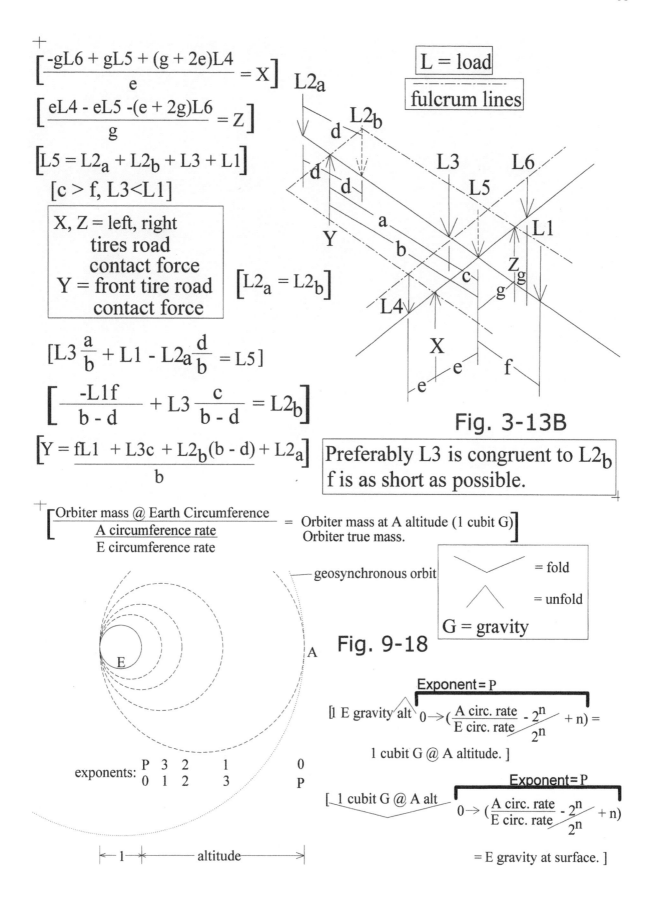

$$\left[\frac{-gL6 + gL5 + (g + 2e)L4}{e} = X\right]$$

$$\left[\frac{eL4 - eL5 -(e + 2g)L6}{g} = Z\right]$$

$$[L5 = L2_a + L2_b + L3 + L1]$$

$$[c > f, L3 < L1]$$

X, Z = left, right
tires road
contact force
Y = front tire road
contact force

$$[L2_a = L2_b]$$

L = load

fulcrum lines

L2a

L2b

L3

L6

L5

L1

Y

a

b

c

Z

g

g

L4

X

e

e

f

e

d d d d

$$\left[L3\frac{a}{b} + L1 - L2a\frac{d}{b} = L5\right]$$

$$\left[\frac{-L1f}{b - d} + L3\frac{c}{b - d} = L2_b\right]$$

$$\left[Y = \frac{fL1 + L3c + L2_b(b - d) + L2_a}{b}\right]$$

Fig. 3-13B

Preferably L3 is congruent to L2$_b$
f is as short as possible.

$$\left[\frac{\frac{\text{Orbiter mass @ Earth Circumference}}{\text{A circumference rate}}}{\text{E circumference rate}} = \begin{array}{l}\text{Orbiter mass at A altitude (1 cubit G)}\\ \text{Orbiter true mass.}\end{array}\right]$$

geosynchronous orbit

= fold

= unfold

G = gravity

Fig. 9-18

E

A

exponents:
$$\begin{array}{ccccc} P & 3 & 2 & 1 & 0 \\ 0 & 1 & 2 & 3 & P \end{array}$$

|← 1 →|←——— altitude ———→|

Exponent = P

$$\left[1\ \text{E gravity alt}\ 0 \to \left(\frac{\frac{\text{A circ. rate}}{\text{E circ. rate}} - 2^n}{2^n} + n\right) = \right.$$

1 cubit G @ A altitude.]

Exponent = P

$$\left[1\ \text{cubit G @ A alt}\ 0 \to \left(\frac{\frac{\text{A circ. rate}}{\text{E circ. rate}} - 2^n}{2^n} + n\right)\right.$$

= E gravity at surface.]

circumference.

The more dense body, traveling at the same velocity as the less dense body, falls behind the less dense body's orbit velocity period by the lengths difference of the two orbits each orbit. This is determined by using a significantly more dense body and sending it on its own orbit from the altitude and velocity of the less dense body.

The eccentricity of an orbit bisected by its mid-orbit altitude describes its density with respect to its "steady" velocity at its prescribed mid-orbit altitude. Its "steady" velocity is its velocity at its mid-orbit altitude.

Density ratio in this case is:

Mid-point between perigee and apogee is the same circumference as the orbit, so halfway between the mid-point and the perigee is the likely orbit.

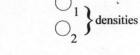

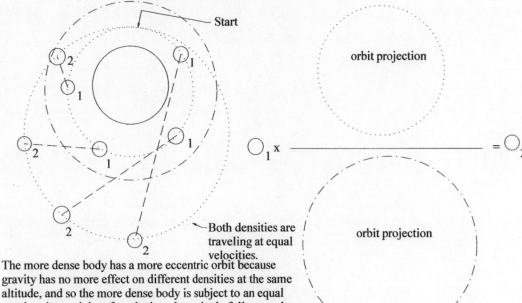

The more dense body has a more eccentric orbit because gravity has no more effect on different densities at the same altitude, and so the more dense body is subject to an equal acceleration and therefore the less dense body follows a closer orbit than the more dense body because less dense matter accelerates faster than more dense matter at the same altitude.

Gravity doesn't change as acceleration occurs. As altitude applies, the angle thrust makes with tangent to the orbit defines gravity. The actual thrust vector does not decrease the angle made with the tangent to the orbit as acceleration velocity reaches y limit at intercept because gravity does not decrease with increasing velocity.

As interval decreases with 1m spacecraft must meet interval time tolerances.

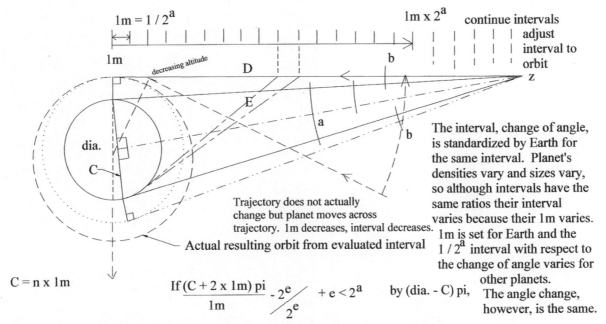

The interval, change of angle, is standardized by Earth for the same interval. Planet's densities vary and sizes vary, so although intervals have the same ratios their interval varies because their 1m varies.

Trajectory does not actually change but planet moves across trajectory. 1m decreases, interval decreases.

Actual resulting orbit from evaluated interval

1m is set for Earth and the $1/2^a$ interval with respect to the change of angle varies for other planets. The angle change, however, is the same.

$C = n \times 1m$

$$\text{If } \frac{(C + 2 \times 1m)\,pi}{1m} - 2^e \Big/ 2^e + e < 2^a \quad \text{by (dia. - C) pi,}$$

then D divided into $1/2^a$ increments intersects E divided into $1/2^a$ increments, and point z may be nearly determined in 1m increments: $1m = 1/2^a$. With respect to (dia - C) pi, the "$< 1/2^a$" increments will have intervals less than the actual $1/2^a$ increment interval. Approximating an interval to displacement time to orbit will require adjustment to the orbit interval displacement time if the alien planet density is undetermined.

When $1m \times 2^a$ orbit circumference increases by one interval or more, the interval decreases by initial interval $\times (1 - (2^a - F/2^a))$, so initial interval aligns 1m on both sides of the planet. F = count of additional intervals from $1/2^a$ goes to $F/2^a$, F can be any number. Calculation of approach interval defines limits of tolerances with respect to 1m @ $1m \times F$ apogee theoretically, actual apogee will be slightly higher. Actual evaluation of $1m \times 1/2^a$ interval is theoretical if the density of the alien celestial body is not known. Theoretically 1m and $1/2^a$ align coefficient with respect to 1e where 1m/e and their $1/2^a$ relationships are constant. Alien celestial body densities will cause variations with respect to intervals.

If altitude above Earth is 1 and orbital circumference is 1_e x $\dfrac{\text{Circumference}}{\text{orbital altitude}}$ $-2^n \Big/ 2^n$ $+ n = 2^a$ x 1_e 7

and duration of orbital period = t, and displacement 1 on orbit is $t/2^a$, Earth's gravity is

$1g$ x $\dfrac{\text{Falling distance}}{\substack{\text{planetary rotational} \\ \text{arc in equal time}}}$ $-2^n \Big/ 2^n$ $+ n$ = Earth's gravity.

| m - moon |
| e - Earth |

Anyway, the moon's altitude $1m$ and the moon's coefficient $1m$ x $\dfrac{\text{orbital altitude of the moon}}{\text{orbit circumference of the moon}}$ $-2^n \Big/ 2^n$ $+ n$

$= 2^{a_m}$ x $1m$.

The moon's gravity was undermined by previous calculations so assume that the moon is an alien celestial body of unknown gravity. The gravity therefore of the moon should be close to ($\dfrac{1m}{1e}$) Earth's gravity

but may vary because of the density of the material out of which the moon is made. If the moon's gravity is ($\dfrac{1m}{1e}$) x Earth's gravity, then the velocity of the spacecraft must be +- accelerated by spacecraft velocity

($\dfrac{1e}{1m}$) x Earth's gravity = moon's orbit velocity.

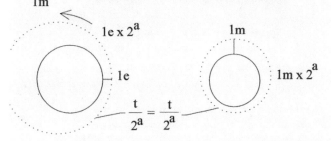

$1m$ x 2 with respect to ($1m$ x 2) x 2^{a-1}

aligns $\dfrac{2t}{2^{a-1}}$ = period, and so forth.

$1m$ x 2 with respect to ($1m$ x 2) 2^a also aligns orbit if $\dfrac{2t}{2^{a-1}}$ = period.

$\dfrac{2t}{2^a}$ = period if $1m$ x any number d aligns with respect to ($1m$ x d) x 2^a

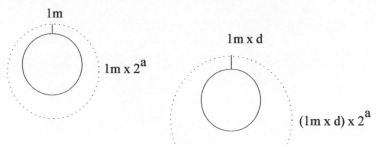

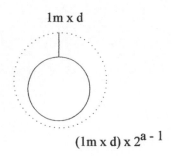

$> 1/2 \ 1/m = b$

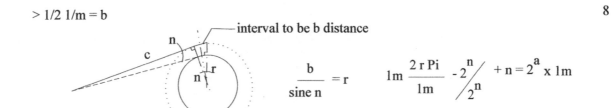

interval to be b distance

$$\frac{b}{\text{sine } n} = r \qquad 1m \frac{2 r \, Pi}{1m} - 2^{n} \Big/ 2^{n} + n = 2^{a} \times 1m$$

Align ($\frac{1m}{1e}$) x Earth's gravity = moon's orbit velocity. With respect to c, change of n should

reflect change of c to be ($\frac{1m}{1e}$) x Earth's gravity = moon's orbit velocity approaching the moon as

c is decreasing with respect to its $1m \times 2^{x}$ coefficient length in $\frac{t}{2^{a}}$ time to $1m$ @ $\frac{t}{2^{a}}$ or its

respective correlative time duration for altitude approaching altitude 1m.

It would seem that if the 1m and the 1e with respect to 2^{a} are the same for both the moon and the Earth then the periods are also the same increment, since the ratio of velocities aligns the period durations.

   Geosynchronous orbit: Entering geosynchronous orbit is not entirely necessary. Enter altitude orbit velocity, +/- accelerate ascend/descend to geosynchronous orbit altitude.

   Enter altitude of celestial body, align Earth's ratio geosynchronous altitude with respect to period at celestial body if possible. Ratio altitude of celestial body with geosynchronous altitude of Earth and calculate values of celestial body variables. Make possible change of altitude, velocity, and period to geosynchronous orbit of celestial body by ratio of Earth period at orbit of celestial body with respect to celestial body change of celestial body rotation rate difference.

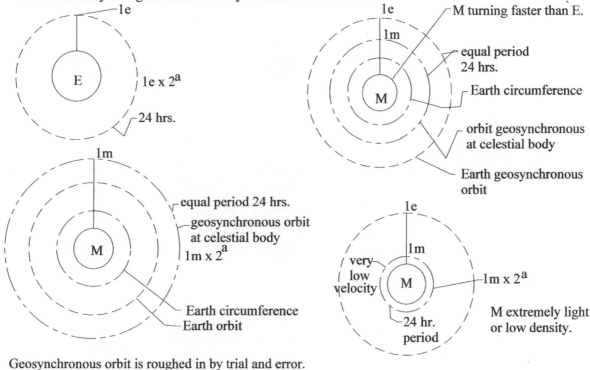

Geosynchronous orbit is roughed in by trial and error.

Otherwise a coefficient with respect to orbit around M may be ascertained by 1m with respect to 1m x 2^a with respect to 1e with respect to 1e x 2^a that's not a geosynchronous orbit, and the same results obtained.

In one case, geosynchronous orbit is obtained by defining the middle altitude between apogee and perigee and calculating the ascent or descent rate from aphihellion or perihelion and calculating the coefficient braking force and time to decelerate the rocket to the middle altitude on ascent to stop ascent/descent on middle altitude while after having put the rocket in orbit around a planetary body at a velocity which sustains an orbit with an apogee and a perigee.

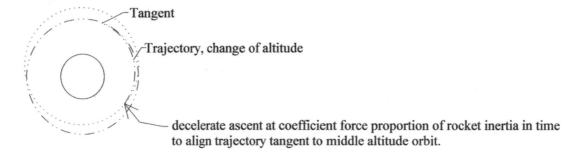

Tangent

Trajectory, change of altitude

decelerate ascent at coefficient force proportion of rocket inertia in time to align trajectory tangent to middle altitude orbit.

Applied coefficient force of rocket inertia moment +/- accelerates the rocket a specific distance in a coefficient time.

As well, the geosynchronous altitude does not have to be the middle altitude and can be above the orbit or beneath the orbit, or can still be inside the orbit but be closer to apogee or perigee depending on the velocity of the spacecraft and the entry altitude.

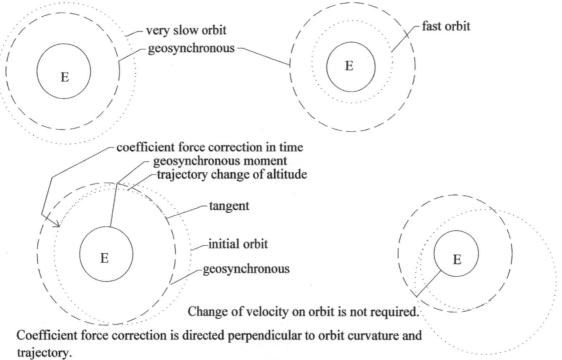

very slow orbit

geosynchronous

fast orbit

coefficient force correction in time

geosynchronous moment

trajectory change of altitude

tangent

initial orbit

geosynchronous

Change of velocity on orbit is not required.

Coefficient force correction is directed perpendicular to orbit curvature and trajectory.

Once alien orbit has been described, discretionary judgement may be determined with respect to what to do with the orbit.

The more dense body, traveling at the same velocity as the less dense body, falls behind the less dense body's orbit velocity period by the lengths difference of the two orbits each orbit. This is determined by using a significantly more dense body and sending it on its own orbit from the altitude and velocity of the less dense body.

The eccentricity of an orbit bisected by its mid-orbit altitude describes its density with respect to its "steady" velocity at its prescribed mid-orbit altitude. Its "steady" velocity is its velocity at its mid-orbit altitude.

Density ratio in this case is:

Mid-point between perigee and apogee is the same circumference as the orbit, so halfway between the mid-point and the perigee is the likely orbit.

Start

orbit projection

densities

Both densities are traveling at equal velocities.

orbit projection

The more dense body has a more eccentric orbit because gravity has no more effect on different densities at the same altitude, and so the more dense body is subject to an equal acceleration and therefore the less dense body follows a closer orbit than the more dense body because less dense matter accelerates faster than more dense matter at the same altitude.

When reentry momentum divided by $Nf \longrightarrow 2^P$, $Nf\,2^P$ is the braking force to inertia at velocity. Only allow $Nf\,2^P$ = operator's weights sum.

P decreases with time a prescribed amount in the allotted time, so N is a calibrated volume with respect to f and speed, to slow the spacecraft once it enters the atmosphere to the desired speed at the desired altitude. Calibration of desired deceleration may be accomplished yet further still, to land in an atmosphere, and a second braking device may be implemented:

$$2^3 \underline{\quad} \overset{2^{3.0125}}{\underline{}} \;\; \widehat{2^P G}\; P, (0 \longrightarrow P)$$

Momentum on descent never exceeds Earth G rate of man momentum.

$$2^2 \underline{\quad} \text{-----------}$$

Engine force = $M\;2^P G\;P, (0 \rightarrow P)$ - operators mass Earthside.

$$2^1 \underline{\quad} \text{--------}$$
$$2^0 \underline{\quad} \text{---------}$$
$$\underline{D} \text{ ----------}$$

$$\underline{} 2^P G$$

$$\boxed{\dfrac{2^P G}{2(0 \rightarrow P)} = \widehat{2^P G}\; P,(0 \rightarrow P)}$$

D = Earth G rate of man momentum distance.

$$D2^{3.0125}$$

At one of the increments $D2^n$ the addition of operators mass $G2^P$ is indexed to apply engine force addition over the $D\,2^n$ distance

in simultaneous time increasing from $G0 \;\rightarrow\; G2^P + M\; \widehat{2^P G}\; P,(0 \rightarrow P)$

$$2^{\left(\frac{\text{Orbital altitude}}{\text{man rate distance Earthside}} - \frac{2^n}{2^n}\right) + n} = 2^{3.0125}, \text{ e.g.}$$

Return from orbit of moon and fall behind Earth's orbit and catch up to
one interval on approach.

This practice is good for planetary return voyages practice
and for commercial tourism to and from the Moon to get
practice for astronauts taking planetary voyages. The practice
must be enter orbited to the Moon, orbit velocity of the Moon
must be eliminated, and return to Earth orbit entry procedure
followed accordingly. This path allows for tangent alignment
of the spacecraft to the planet without the planet's trajectory
intersecting the trajectory of the spacecraft on entry although
the spacecraft is required to stop in space and accelerate to
enter orbit. Alignment of tangent entry trajectory will be routine
after some practice.

M

1m

E

16
braking
8
4
2
1
1/2
1/4 1/4

4

16

8

4
2
1/2
1/4 1/4

2

1
1/4 1/2
1/4

16 1/2 1/4

2

1 x 2^0
1 x 2^{-1}
1/4
1/4

Curvatures

reverse to brake

velocity impact force
= 2Msm/v @ 1 x 2^1

Rocket thrust at braking = 1Msm/v + operators.
Rocket thrust may equal $\frac{1}{2^n}$ Msm/v + operators.

[Msm/v = spacecraft weight, m/v = model/vessel, meaning a or any particular
type of craft.]

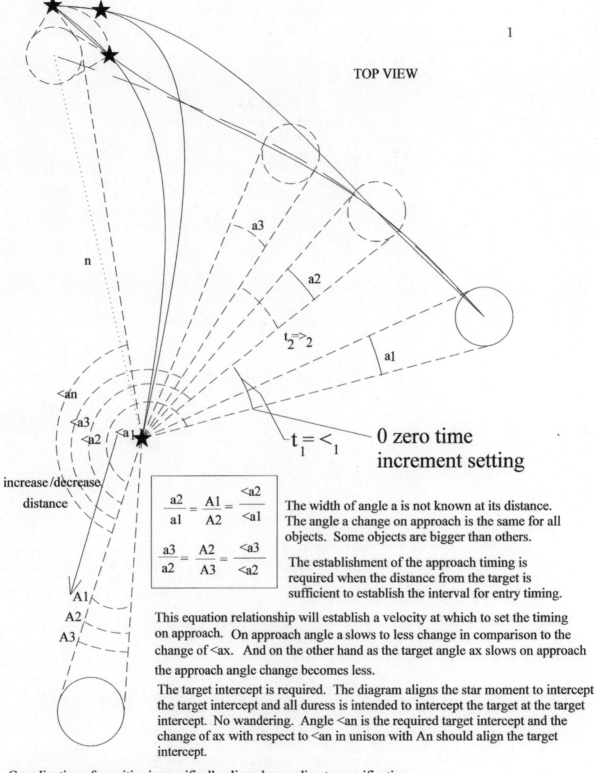

TOP VIEW

1

a3

a2

$t_2 = <_2$

a1

n

<an

<a3

<a2 <a1

$t_1 = <_1$

0 zero time
increment setting

increase /decrease
distance

$$\frac{a2}{a1} = \frac{A1}{A2} = \frac{<a2}{<a1}$$

$$\frac{a3}{a2} = \frac{A2}{A3} = \frac{<a3}{<a2}$$

A1

A2

A3

The width of angle a is not known at its distance.
The angle a change on approach is the same for all
objects. Some objects are bigger than others.

The establishment of the approach timing is
required when the distance from the target is
sufficient to establish the interval for entry timing.

This equation relationship will establish a velocity at which to set the timing
on approach. On approach angle a slows to less change in comparison to the
change of <ax. And on the other hand as the target angle ax slows on approach
the approach angle change becomes less.

The target intercept is required. The diagram aligns the star moment to intercept
the target intercept and all duress is intended to intercept the target at the target
intercept. No wandering. Angle <an is the required target intercept and the
change of ax with respect to <an in unison with An should align the target
intercept.

Coordination of gravities is specifically aligned according to specifications.

Orbital entry off the ecliptic will be hypercyclonic.

Theory of the Sound Barrier:

air displacement of aircraft volume = V cubic units

the aircraft moves one unit volume and displaces its displacement $\frac{1}{p} = 1v$ cubic volume

N (air density) is specific at altitude.

f (aircraft displacement volume) is constant.

$\underline{B}$ weight of the aircraft
p fold exponent after calculation

$$p \times (1v = Vp\frac{1}{p} = V) = \frac{B}{p} \quad \frac{\text{weight of aircraft}}{p} = \text{weight of displaced air at altitude.}$$

$$\frac{B}{p} < p \times (1v = Vp\frac{1}{p} = V)$$

$$\frac{\text{weight of aircraft}}{p} < \text{weight of displaced air on the ground.}$$

Otherwise, eliminate p in the denominator of B/p, and raise all other p's as base two exponents. The swept volume becomes $1v2^p = B$.

A x small angle sweep / larger angle sweep = %A.

Assuming that one of the angles of the trapezoid is known and that B is less than A, then f angle is 1/2 more than the right side of the trapezoid and 1/2 less than the left side of the trapezoid. Or vice versa.

So if the right side of the trapezoid is 60 degrees and the left side of the trapezoid is 64 degrees then f is 62 degrees. Nevertheless, the height of f must be known, and two of the angles of the triangle.

The previous calculations for the trapezoid are true. Now the equation fg = area is true as well, as is fg = volume when g is an area.

It is also true when subtracting areas times their f fractional height length from fg to equal the difference at the wing's ends when finding a B-leading volume addition. Also the f fraction of length times its area equaling the wing's subsegment volume in addition to the B-trailing volume can also be found.

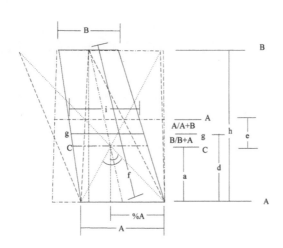

This is the formula for finding the center of
an oblique trapezoid, eg: a wing.

However, it has since been discovered that wings, e.g. the delta wing, is balanced by the aA/b = B method although the center of the delta wing volume geometry is different from its aA/b = B balance equilibrium moment. Swept back delta wings are still balanced by the aA/b = B formula although for symmetry it is easier to make the plane fly straight if it looks the same on the other side and therefore even if the wings are swept back and the delta seems more difficult to figure out, finding the aA/b = B equilibrium moment for the geometry of the wingS will put the aA/b = B cantilever in the center of the airplane, and where the formula balances is where the delta wings geometry bisector plane orbits with the wind. If the airplane (vessel, vehicle, etc.) has multiple delta airfoils all their individual aA/b = B fulcrums are individually found and then all together their Em = 0 is where the delta air foil lift is balanced. Including the remaining geometry in the aircraft body volume and all of the ups and downs forces throughout the geometric displacement of the model and including the delta Em = 0, again find the remaining Em = 0 and continue to further the resolution of details involving balancing the model.

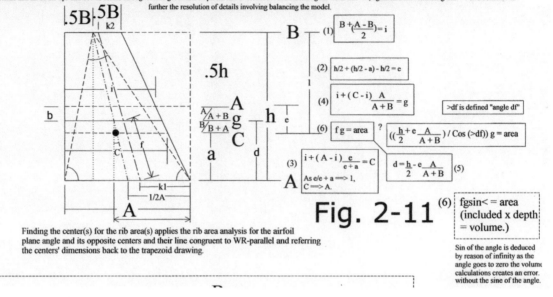

Fig. 2-11

Finding the center(s) for the rib area(s) applies the rib area analysis for the airfoil plane angle and its opposite centers and their line congruent to WR-parallel and referring the centers' dimensions back to the trapezoid drawing.

(6) fgsin< = area (included x depth = volume.)

Sin of the angle is deduced by reason of infinity as the angle goes to zero the volume calculations creates an error without the sine of the angle.

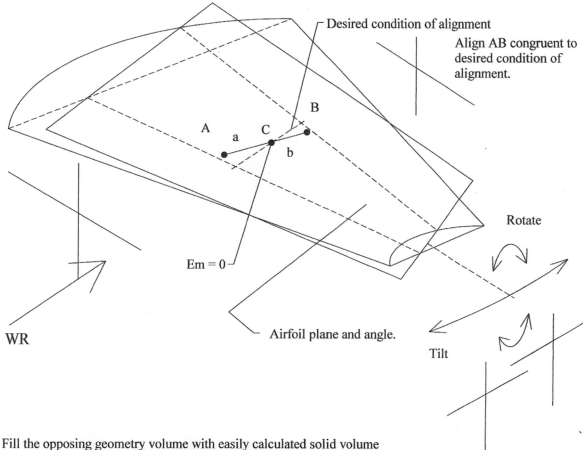

Desired condition of alignment

Align AB congruent to desired condition of alignment.

B

A   C
  a
       b

Em = 0

Rotate

Airfoil plane and angle.

WR

Tilt

Fill the opposing geometry volume with easily calculated solid volume shapes until the desired geometric shape is described and the fulcrum defines the Em = 0 for A and B such that rotational alignment of the airfoil plane and static balancing calculations describe ▬▬▬▬▬▬▬▬▬▬ $=(a + b)A \diagup (A + B) = B$ congruent to WR.

Repetitive recalculation is required during the rotational alignment of the airfoil plane alignment such that the input of subvolume shapes may be reapplied. This repetitive recalculation is the basis on which the necessity to incorporate the use of software and a computer eliminates paperwork. Static equations and the input of Right and equilateral and obtuse geometrical solids which volumes can be easily calculated makes a computer simplify the task of repetitive calculations.

base equations:

$$X^Y$$

$$0 \quad X^0 \ X^1 \quad X^2 \qquad X^n \qquad \frac{B}{Nf} \qquad X^{n+1}$$

$$\left[J^{\left(\frac{\frac{B}{Nf} - J^m}{J^{m+1} - J^m} + m\right)} Nf = J^L Nf \text{ lbs.} = B \text{ lbs.} \right]$$

$$J^L \text{ lbs.} = \text{pounds.}$$

$$\left[A^{\left(\frac{\frac{J^L}{Nf} - A^n}{A^{n+1} - A^n} + n\right)} = A^R Nf, \text{ change of base.} \quad J^L \rightarrow A^R Nf \right]$$

$$\boxed{N = \text{density}, \ f = \text{volume}}$$

Other bases

$$\left[2^{\left(\frac{1000 - 2^9}{2^{9+1} - 2^9} + 9\right)} = 7^{\left(\frac{1000 - 7^3}{7^{3+1} - 7^3} + 3\right)} \right.$$

$$= 7^{3.31924} = 2^{9.95313} = 1000. \right]$$

Obviously $\dfrac{\log 1000}{\log 2} \neq 9.95313$, and

$\dfrac{\log 1000}{\log 7} \neq 3.31924$ But, on the calculator:

$2^{9.95313} = 991.267$, and $7^{3.31924} = 638.386$
Something is amiss.

base equations:

$$x^y$$

$$0 \quad x^0 \quad x^1 \quad x^2 \quad \cdots \quad x^n \quad \frac{B}{Nf} \quad x^{n+1}$$

$$\left[J^{\left(\frac{\frac{B}{Nf} - J^m}{J^{m+1} - J^m} + m\right)}{}_{Nf} = J^L{}_{Nf} \text{ lbs.} = B \text{ lbs.} \right]$$

$$\left[A^{\left(\frac{\frac{J^L}{Nf} - A^n}{A^{n+1} - A^n} + n\right)} = A^R{}_{Nf}, \text{ change of base.} \quad J^L \rightarrow A^R{}_{Nf} \right]$$

lbs. = pounds.

$\boxed{N = \text{density}, \ f = \text{volume}}$

Other bases

$$\left[2^{\left(\frac{1000 - 2^9}{2^{9+1} - 2^9} + 9\right)} = 7^{\left(\frac{1000 - 7^3}{7^{3+1} - 7^3} + 3\right)} \right.$$

$$= 7^{3.31924} = 2^{9.95313} = 1000. \left. \right]$$

Obviously $\dfrac{\log 1000}{\log 2} \neq 9.95313$, and

$\dfrac{\log 1000}{\log 7} \neq 3.31924$ But, on the calculator:

$2^{9.95313} = 991.267$, and $7^{3.31924} = 638.386$
Something is amiss.

$$\left[\dfrac{2^{\left(\dfrac{\dfrac{BB}{\text{Cos} <1 \,\text{Cos} <2}}{Nf} - 2^n\right)}}{2^n} {}^{-n)} Nf = Nf\, 2^{(p)} \right.$$

= elastic load
on Nf. along $N f 2^q$

2^q is the thrust line vector $Nf2^p$ is parrallel to.

In reverse:

$$\left[2^{(p)} = \dfrac{\cancel{Nf}\,\cancel{2}}{\cancel{Nf}}\, {}^{(p - n\,(2^{n+1} - 2^n) + 2^n\,(\text{Cos} <2\,\text{Cos} <1)} \right]$$

= elastic force applied to Nf.

$$\left[2^{n+1} - 2^n = 2^n \right] \qquad Nf2^p = Nf2^q \quad p \neq q \text{ when } f(p) \neq f(q).$$

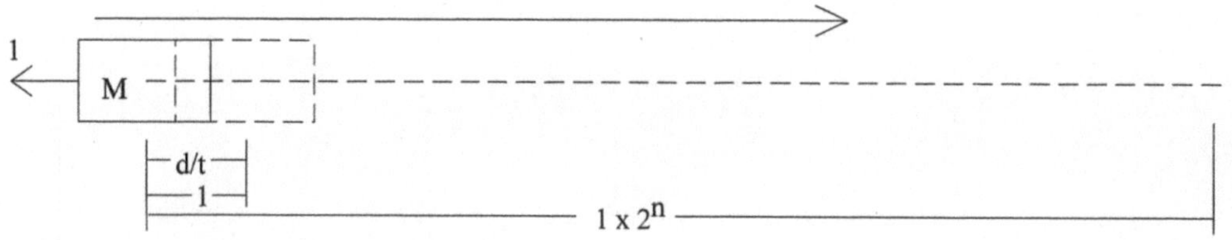

+

d/t = one momentum mass force.

slug = momentum impact reaction force equal to mass of density.

time increment = 1 standard Earth second.

Fig. 9-12

PROBLEM: If a mass of one unit mass displaces its weight force momentum at 1 distance unit in one time increment, and has 1×2^n displaced distance in one time increment, and a one unit force equal to the mass of the density acts in the opposite direction to brake the momentum to a stop: how many seconds will stop the momentum and how far will the momentum travel simultaneously?

ANSWER: Assume mass of momentum is one pound and momentum distance force is one foot in one second at one pound. Assume rate is 2^4 feet per second = 16 feet per second. Then, 1 pound brakes 16 pounds momentum at deceleration from 2^4 to 0 : 16 + 8 + 4 + 2 + 1 and one foot from 1 to zero = 32 feet stopping distance, and there are five plus signs, so five seconds will stop the momentum.

PROBLEM: At what rate is a mass traveling when its impact reaction force is equal to its weight?
SOLUTION: unknown.

+

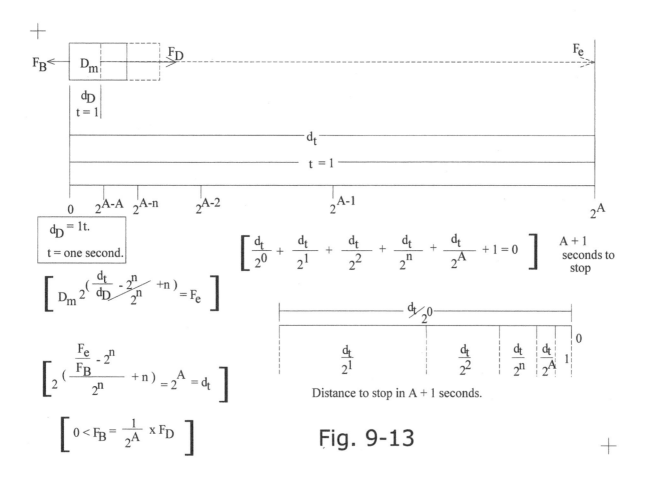

$$\left[D_m 2^{\left(\frac{d_t}{d_D} - \frac{2^n}{2^n} + n \right)} = F_e \right]$$

$$\left[\frac{d_t}{2^0} + \frac{d_t}{2^1} + \frac{d_t}{2^2} + \frac{d_t}{2^n} + \frac{d_t}{2^A} + 1 = 0 \right] \quad \begin{array}{l} A+1 \\ \text{seconds to} \\ \text{stop} \end{array}$$

$$\left[2^{\left(\frac{\frac{F_e}{F_B} - 2^n}{2^n} + n \right)} = 2^A = d_t \right]$$

Distance to stop in A + 1 seconds.

$$\left[0 < F_B = \frac{1}{2^A} \times F_D \right]$$

Fig. 9-13

METEOR INCINERATION EQUATION (simplicity of base two)

$$2^{\left(\frac{\left(\frac{\text{impact reaction force}}{\text{weight of the meteor}} + \frac{\text{weight of the meteor}}{Nf \text{ of the meteor}} + \frac{\text{impact reaction force}}{Nf \text{ of the meteor}} \right) - 2^n}{2^n} \right) + n} Nf =$$

applied reaction force to the meteor of the surrounding air.

The change in temperature may also be the same $2^p t_{air @ N}$. This is the largest value for 2^p there is. Aligning the impact reaction force with respect to the change in temperature will derive a velocity with respect to the meteor. Note that the meteor "rate" is a velocity and therefore imparts no elapsed time to the meteor on its trajectory. The equation takes place in zero time and all the applied variables are calculated simultaneously. Changes of altitude vary N. f changes as the meteor is consumed. Air acts on the entire volume of the meteor consuming the meteor as a body, not simply from the leading subsegment. As can be seen from video a meteor is encompassed by the atmosphere a column of fire streams directly trailing from the center aft of the meteor.

To align p so that $t_{air @ N} 2^p$ will burn up a meteor align the impact reaction force (in pounds) at its respective velocity until $t_{air @ N} 2^p$ will incinerate the material out of which a meteor is made, in a fraction of a second.

t = temperature, N = air density at altitude, f = displacement volume of subject body, p is fold force multiple. force units are in pounds, N is in lbs/cu.ft., f is in cu.ft.

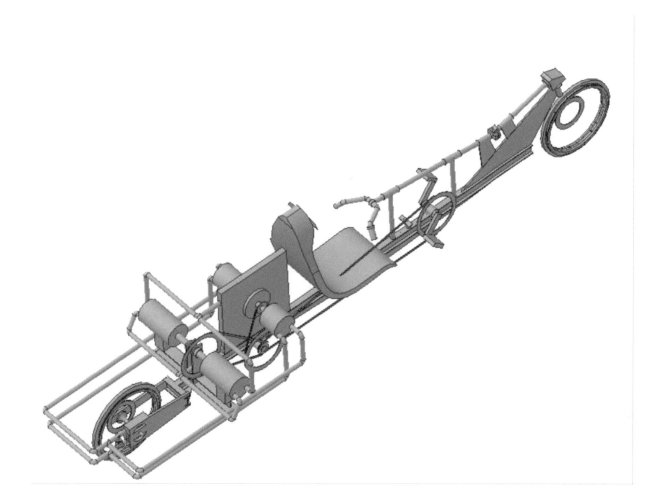

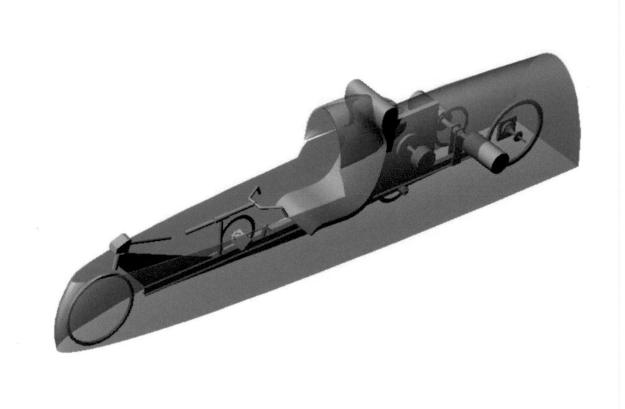

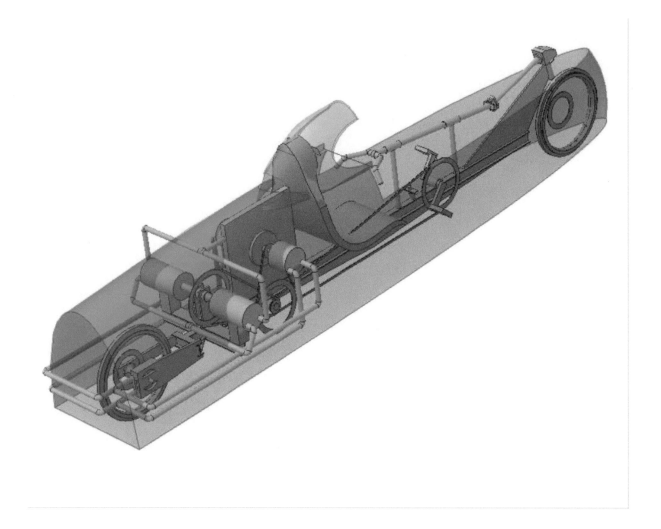

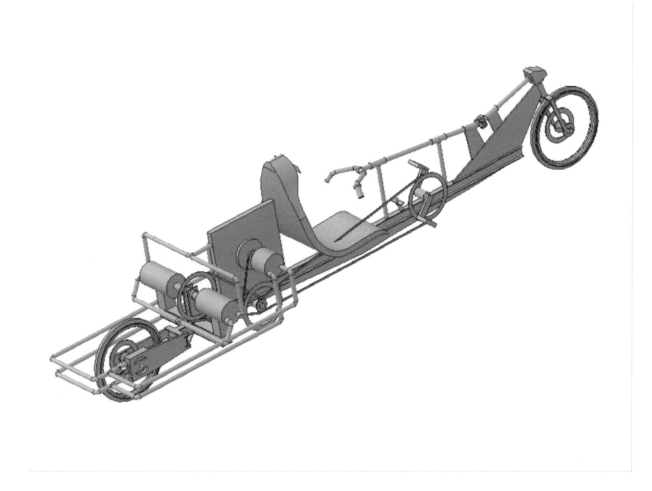

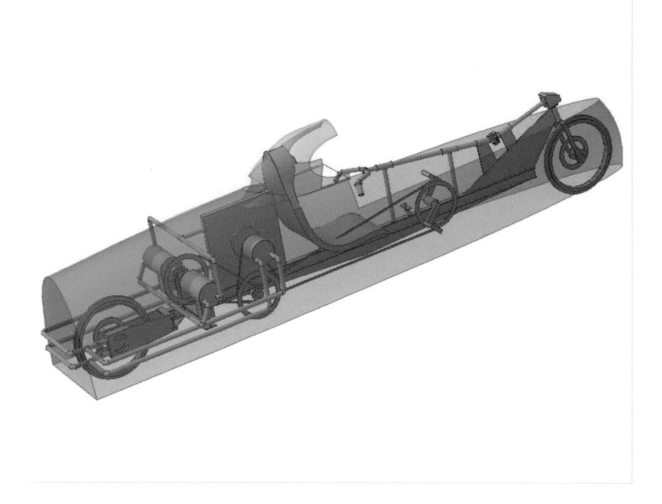

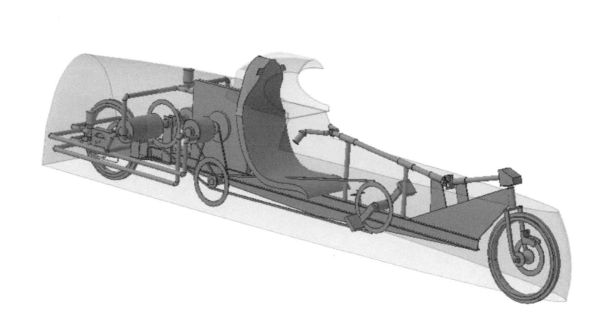

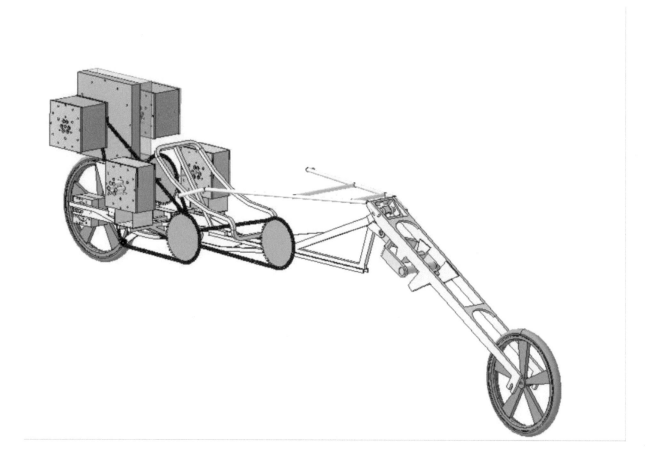

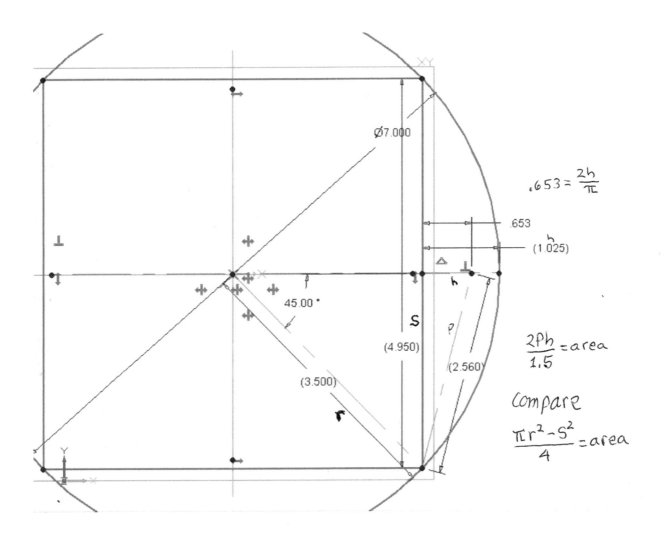

Ø7.000

$.653 = \dfrac{2h}{\pi}$

.653

$\begin{array}{c} h \\ (1.025) \end{array}$

45.00°

S

(4.950)

(3.500)

r

(2.560)

$\dfrac{2Ph}{1.5} = area$

Compare

$\dfrac{\pi r^2 - S^2}{4} = area$

I discovered this formula for calculating the area of a chord on August 18 2011.

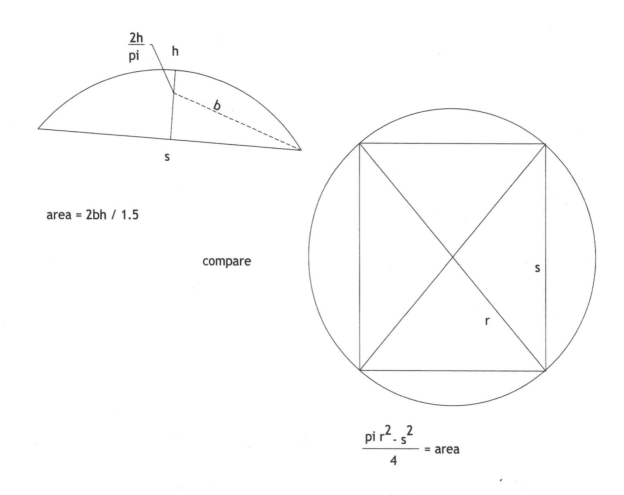

$$\frac{2h}{pi}$$ h

b

s

area = 2bh / 1.5

compare

s

r

$$\frac{pi \, r^2 - s^2}{4} = area$$

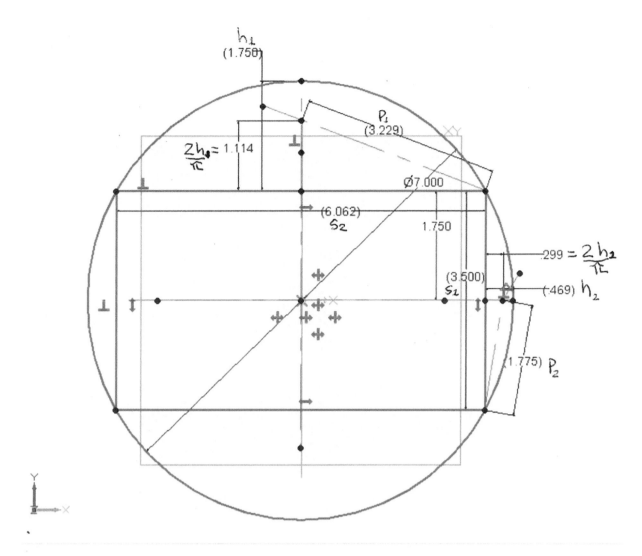

$$\pi r^2 - S_1 \times S_2 = 17.268\ ^2in$$

$$\frac{2P_1h_1}{1.5} + \frac{2P_2h_2}{1.5} = 17.289\ ^2in$$

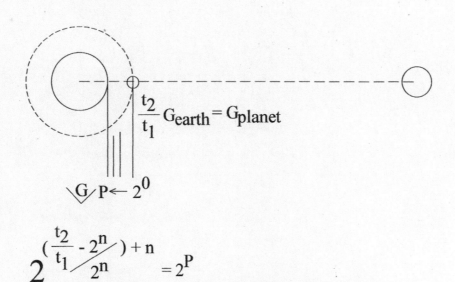

$$\frac{t_2}{t_1} G_{earth} = G_{planet}$$

$$G \swarrow P \leftarrow 2^0$$

$$2^{\left(\frac{\frac{t_2}{t_1} - 2^n}{2^n}\right) + n} = 2^P$$

t_1 = Earth orbit duration at geosynchronous altitude

t_2 = Planetside orbit duration at geosynchronous altitude

Apply descent rate initially at the operator's momentum, equal to one (1) Earth applied rate of gravity. Apply engine braking force to sustain constant rate on descent as $0 \longrightarrow 2^P G$. Remaining altitude is $\frac{t_2}{t_1}$ x operator's momentum

1 rate distance in equal time, and engine force maintaining 1 Earth G rate to operators mass x 2^P by increments of $\frac{\text{altitude remaining}}{2^{n_1}}$ n_1 in one second.

Otherwise, to increase the time multiply altitude by 2^{n_2} and increase the applied engine force increase time by 2^{n_2}.

Multiply $\frac{t_2}{t_1}$ x planet orbit circumference to align gravitational force equal to Earth geosynchronous orbit Gravity force at planet.
This new altitude will experience safe assembly gravity effects to the spacecraft.

Depending on the severity of $G \swarrow P$ addition of operator's force x 2^P over altitude (2^P), as $2^P \swarrow P \searrow P$ aligns rate to Earth rate of operator's

momentum 1 rate distance in equal time, the applied force to the operators goes to operator's x 2^P at t_2 altitude. Go to t_2 circumference x $\frac{t_2}{t_1}$

altitude reducing applied acceleration force to operators x $2^P \searrow P$ time.

When reentry momentum divided by Nf $\longrightarrow 2^P$, Nf 2^P is the braking force to inertia at velocity. Only allow Nf $2^P=$ operator's weights sum.

P decreases with time a prescribed amount in the allotted time, so N is a calibrated volume with respect to f and speed, to slow the spacecraft once it enters the atmosphere to the desired speed at the desired altitude. Calibration of desired deceleration may be accomplished yet further still, to land in an atmosphere, and a second braking device may be implemented:

$2^3 \underline{\quad} \quad 2^{2^{3.0125}} \text{-------} \overset{\frown}{2^P G} \; P, (0 \longrightarrow P)$

Momentum on descent never exceeds Earth G rate of man momentum.

Engine force $= M \; 2^P G \; P, (0 \rightarrow P)$ - operators mass Earthside.

$2^2 \underline{\quad} \text{--------}$

$2^1 \underline{\quad} \text{------}$
$2^0 \underline{\quad}$
$\quad D \text{------}$

$\qquad 2^P G$

$$\frac{2^P G}{2(0 \rightarrow P)} = \overset{\frown}{2^P G} \; P, (0 \rightarrow P)$$

D = Earth G rate of man momentum distance.

$D2^{3.0125}$ At one of the increments $D2^n$ the addition of operators mas $G2^P$ is indexed to apply engine force addition over the $D \; 2^n$ distance

in simultaneous time increasing from $G0 \rightarrow G2^P + M \; \overset{\frown}{2^P G} \; P, (0 \rightarrow P)$

$$2^{\left(\frac{\text{Orbital altitude}}{\text{man rate distance Earthside}} \cdot \frac{-2^n}{2^n}\right)+n} = 2^{3.0125}, \text{ e.g.}$$

in t_2 circumference radius x $\dfrac{t_2}{t_1}$ altitude - t_2 radius

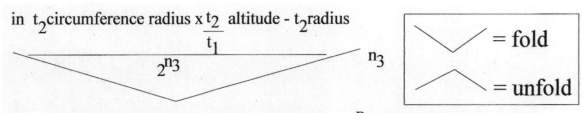

2^{n_3} n_3

It may be behooving to maintain operator's 2^P beyond t_2 circumference to t_2 circumference $\dfrac{t_2}{t_1}$ = altitude, and accelerate tangentially to new altitude.

Also, the indexing of curvature on trajectory to tangent to t_2 circumference x $\dfrac{t_2}{t_1}$ altitude = new altitude, is also indexed with respect to the momentum on ascent. Find altitude on ascent:

$$\dfrac{\text{Operator's mass Earthside}^P}{2^P} \longrightarrow 2^P G @ t_2$$

Calculating pitch indexing angle with perpendicular distance to New altitude and time duration unfold increasing simultaneously with decreasing distance:

Apply seat of the pants to manage altitude difference in $2t$ x times to A.

t/x corresponds to unfold x.

Sp[acecraft must carry reserves spares empty to transfer excess lander fuel etc. upon returning to orbit.

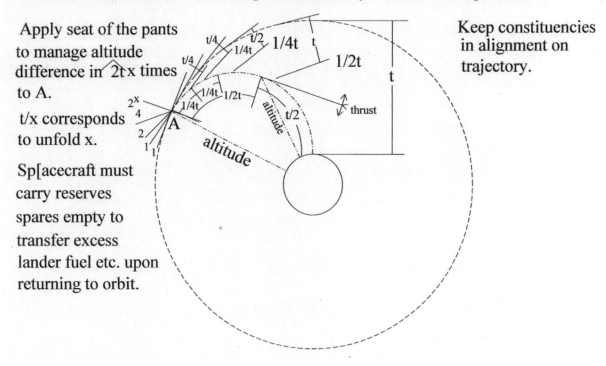

Keep constituencies in alignment on trajectory.

Instructions:
Drivetrain

Construct the drive train assembly.
Construct the frame.
Parts precision: NO tolerance (± 0.000)
Contacting surfaces finished the same.
All Reactive Control Frame variables to
no tolerance.
Contacting surfaces roughness to be perfectly
parallel, concentric, perpendicular

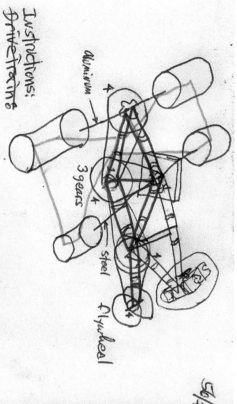

Aluminium

3 gears

steel   flywheel

56/44

Bearings: dimensions, tolerances

$$6.3^2 \times \pi \times .5 - 6^2 \pi \cdot '$$
$$= 78.543 \text{ in}$$
$$\frac{6.99^2 \text{ in}}{} = 5.61$$

$$\frac{12.777 \text{ in}}{5.618 \text{ in}} = 2.156$$

$$2.156 \times 8 = 17.24 \text{A}$$
one revolution.

$$\frac{5.618 \text{ in}}{17.246} \rightarrow 32.63 \text{in}$$

$$.5 \text{in} (17\pi = 6.5 \frac{3}{2}\pi)$$

$$\sqrt{\frac{4.554^3 \text{in}}{.5 \text{in}} + \frac{6.5^2 \pi}{\pi}} =$$

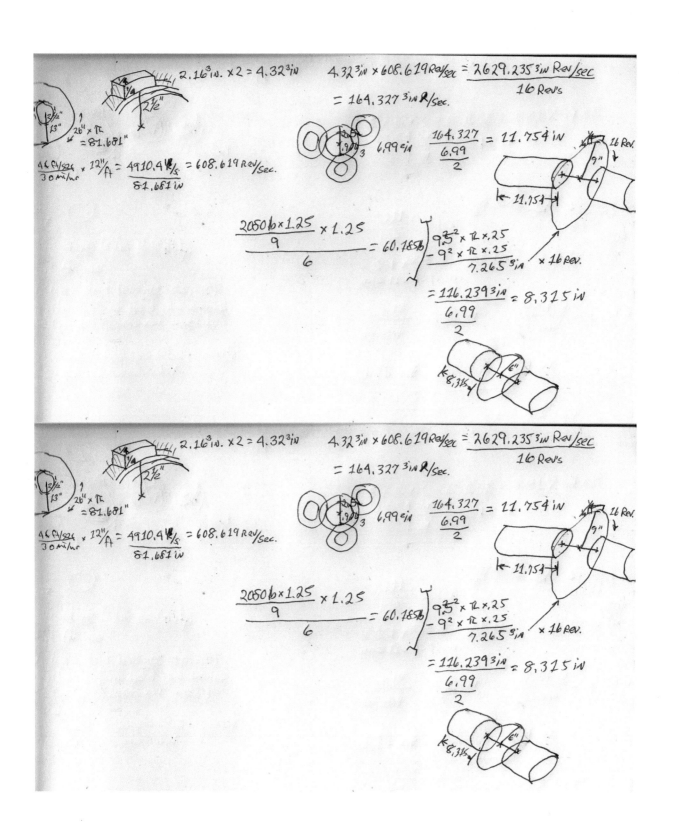

97

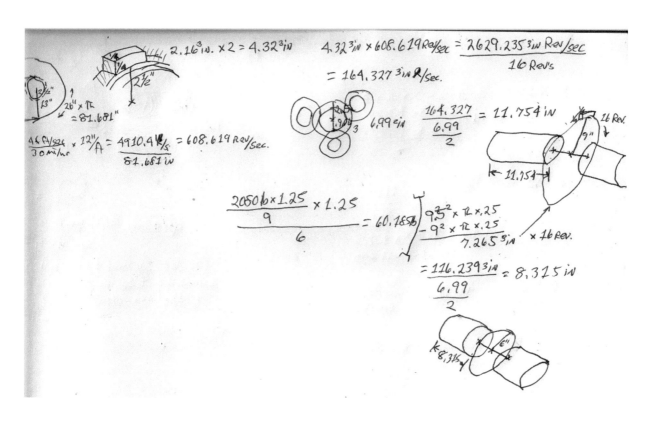

$2.16^3 in. \times 2 = 4.32^3 in$

$4.32^3 in \times 608.619 \, Rev/sec = \dfrac{2629.235^3 in \, Rev/sec}{16 \, Rev's}$

$= 164.327^3 in \, R/sec.$

$13\tfrac{1}{2}"$ $13"$

$26" \times \pi = 81.681"$

$\dfrac{46 \, ft/sec}{30 \, mi/hr} \times 12"/ft = \dfrac{4910.4 \, in/s}{81.681 \, in} = 608.619 \, Rev/sec.$

3.5 1.909 3 $6.99 \, in$

$\dfrac{164.327}{\frac{6.99}{2}} = 11.754 \, in$

16 Rev. $9"$ $11.754"$

$\dfrac{2050 \, b \times 1.25}{9} \times 1.25 = 60.185b$

$\left.\begin{array}{l} 9\tfrac{5}{8}^2 \times \pi \times .25 \\ - \, 9^2 \times \pi \times .25 \\ \hline 7.265^3 in \end{array}\right\} \times 16 \, Rev.$

$= \dfrac{116.239^3 in}{\frac{6.99}{2}} = 8.315 \, in$

8.315 $6"$

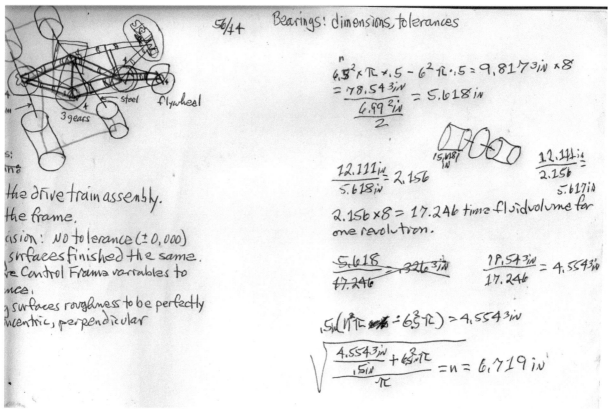

$56/44$

Bearings: dimensions, tolerances

the drive train assembly.
the frame.
...ision: No tolerance ($\pm 0,000$)
...surfaces finished the same.
...e Control Frame variables to
...nce.
...g surfaces roughness to be perfectly
...ncentric, perpendicular

3 gears   steel   flywheel

$6.5^2 \times \pi \times .5 - 6^2 \pi \times .5 = 9.817^3 in \times 8$

$= 78.543^3 in = 5.618 in$

$\dfrac{6.99^2 in}{2}$

$15.981 \, in$

$\dfrac{12.111 \, in}{5.618 \, in} = 2.156$

$\dfrac{12.141 in}{2.156} = 5.617 in$

$2.156 \times 8 = 17.246$ times fluid volume for one revolution.

$\dfrac{5.618}{17.246} = 326^3 in$

$\dfrac{78.543 \, in}{17.246} = 4.5543 \, in$

$.5 in (n^2 \pi - 6.5^2 \pi) = 4.554^3 in$

$\sqrt{\dfrac{\frac{4.5543 in}{.5 in} + 6.5^3 in \pi}{\pi}} = n = 6.719 \, in$

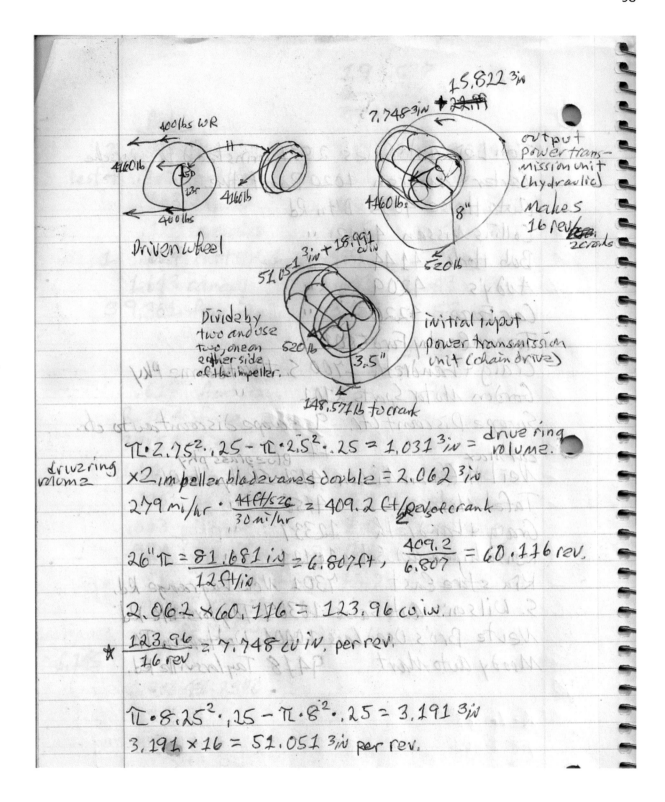

400 lbs WR

15.822^{3in} + ~~2.2 in~~

7.748^{3in}

416 lbs

416 lbs

400 lbs

Driven wheel

output power transmission unit (hydraulic)

Makes 16 rev/crank, 2 cranks

416 lbs

8"

520 lb

51.051^{3in} + 15.991 cu in

Divide by two and use two, one on 2 other side of the impeller.

520 lb

3.5"

initial input power transmission unit (chain drive)

148.571 lb to crank

driving ring volume

$\pi \cdot 2.75^2 \cdot .25 - \pi \cdot 2.5^2 \cdot .25 = 1.031^{3in} =$ driving ring volume.

$\times 2$ impeller blade vanes double $= 2.062^{3in}$

279 mi/hr $\cdot \dfrac{44 ft/sec}{30 mi/hr} = 409.2$ ft/rev of crank

$26" \pi = \dfrac{81.681 in}{12 ft/in} = 6.807 ft$, $\dfrac{409.2}{6.807} = 60.776$ rev.

$2.062 \times 60.776 = 123.96$ cu in.

$\bigstar \dfrac{123.96}{16 rev} = 7.748$ cu in. per rev.

$\pi \cdot 8.25^2 \cdot .25 - \pi \cdot 8^2 \cdot .25 = 3.191^{3in}$

$3.191 \times 16 = 51.051^{3in}$ per rev.

$$\frac{\frac{4160\,lb}{53.9\,^{lb}/_{3A} \cdot .889\,^{3}in}}{12^{3\,3in}/_{3A}} - \frac{2^{n}\,^{17}}{2^{n}} + n = P = \frac{18.143}{17.145}$$

$$(\pi \cdot 2.843^{2\,.781} \cdot .125^{.032}_{.0625} - 2.5^{2}\pi \cdot .125^{.032}_{.0625})2 + \pi \cdot 2.843^{2\,.781} \cdot .25 - \pi \cdot$$
$$2.25^{2} \cdot .25 = 3.543\,^{3}in \quad \frac{.889}{2}\,^{3}in = .445$$

$$P \cdot .445 = 8.074\,^{3}in$$
$$P \cdot .889\,^{3}in = 15.242\,^{3}in /_{2} =$$
$$15.242 + 7.748 = 22.99\,^{3}in \quad 15.822\,^{3}in$$
$$8.074$$

$$\frac{\frac{520\,lb}{53.9 \cdot 1.4133\,in}}{12^{3}} - \frac{2^{n}\,^{13}}{2^{n}} + n = P = 13.44$$

$$(\pi \cdot 8.373^{2\,287} \cdot .0625^{.032} - \pi \cdot 8^{2} \cdot .0625^{.032})2 + \pi \cdot 8.373^{2\,281} \cdot .25 -$$
$$\pi \cdot 8.25^{2} \cdot .25 = 6.015 \quad \frac{2.825}{2} = 1.413\,^{3}in$$

46.794
22.6 ³in

$$P \cdot 1.413 = 18.991\,^{3}in$$

clearance volume = initial unloaded clearance volume × P

This is the formula for finding the center of an oblique trapezoid, eg: a wing.

However, it has since been discovered that wings, e.g. the delta wing, is balanced by the aA/b = B method although the center of the delta wing volume geometry is different from its aA/b = B balance equilibrium moment. Swept back delta wings are still balanced by the aA/b = B formula although for symmetry it is easier to make the plane fly straight if it looks the same on the other side and therefore even if the wings are swept back and the delta seems more difficult to figure out, finding the aA/b = B equilibrium moment for the geometry of the wingS will put the aA/b = B cantilever in the center of the airplane, and where the formula balances is where the delta wings geometry bisector plane orbits with the wind. If the airplane (vessel, vehicle, etc.) has multiple delta airfoils all their individual aA/b = B fulcrums are individually found and then all together their Em = 0 is where the delta air foil lift is balanced. Including the remaining geometry in the aircraft body volume and all of the ups and downs forces throughout the geometric displacement of the model and including the delta Em = 0, again find the remaining Em = 0 and continue to further the resolution of details involving balancing the model.

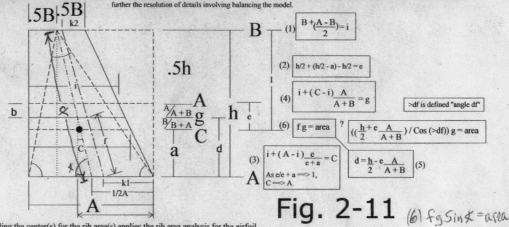

Fig. 2-11 (6) fg Sin ∠ = area

Finding the center(s) for the rib area(s) applies the rib area analysis for the airfoil plane angle and its opposite centers and their line congruent to WR-parallel and referring the centers' dimensions back to the trapezoid drawing.

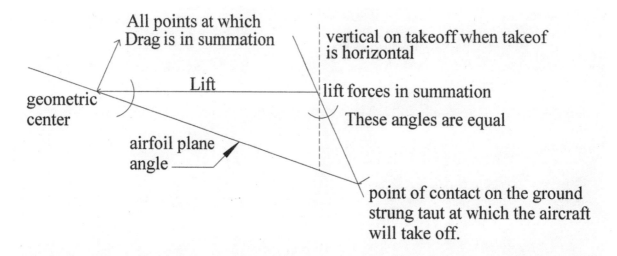

This is perhaps the minimum at which the drag will be sufficient to lift the aircraft provided the aircraft is designed to have the drag to weight ratio to be conformed to this sine ratio.

Notice this strung taut point may be applied even if the aircraft is below ground since the point will be strung taut and thus the aircraft will climb, as if it were on a string. A more accurate kind of "string" is a rubber band.

This is the formula for finding the center of an oblique trapezoid, eg: a wing.

However, it has since been discovered that wings, e.g. the delta wing, is balanced by the aA/b = B method although the center of the delta wing volume geometry is different from its aA/b = B balance equilibrium moment. Swept back delta wings are still balanced by the aA/b = B formula although for symmetry it is easier to make the plane fly straight if it looks the same on the other side and therefore even if the wings are swept back and the delta seems more difficult to figure out, finding the aA/b = B equilibrium moment for the geometry of the wingS will put the aA/b = B cantilever in the center of the airplane, and where the formula balances is where the delta wings geometry bisector plane orbits with the wind. If the airplane (vessel, vehicle, etc.) has multiple delta airfoils all their individual aA/b = B fulcrums are individually found and then all together their Em = 0 is where the delta air foil lift is balanced. Including the remaining geometry in the aircraft body volume and all of the ups and downs forces throughout the geometric displacement of the model and including the delta Em = 0, again find the remaining Em = 0 and continue to further the resolution of details involving balancing the model.

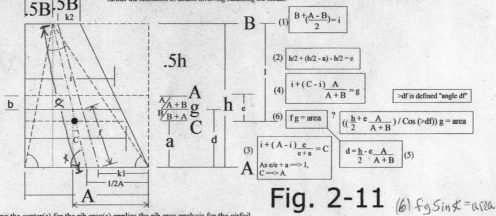

Finding the center(s) for the rib area(s) applies the rib area analysis for the airfoil plane angle and its opposite centers and their line congruent to WR-parallel and referring the centers' dimensions back to the trapezoid drawing.

Fig. 2-11

$$\left[\frac{B}{A}(A - (\frac{h}{h} - \frac{a}{h})A = \frac{B}{A}(A - C) = \text{ the difference of } g - C. \right]$$

$$\left[\frac{B}{A}(A-(\frac{h}{h} - \frac{a}{h})A) + ((\frac{h}{h} - \frac{a}{h})A = \right.$$

$$\left. \frac{B}{A}(A - C) + C = g. \right]$$

$$[gh = \text{volume when A, B, C, and g are areas.}]$$

$$\left[(h - .5h - a)\frac{C}{A} = b \right] \qquad [.5h - b = d]$$

$$\left[(\frac{h}{h} - \frac{a}{h})A = C \right]$$ Original works left for speculation

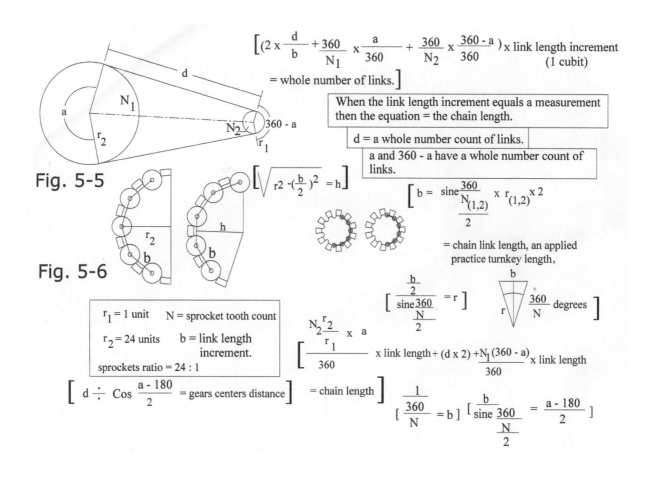

$$\left[(2 \times \frac{d}{b} + \frac{360}{N_1} \times \frac{a}{360} + \frac{360}{N_2} \times \frac{360 - a}{360}) \times \text{link length increment} \right.$$
$$(1 \text{ cubit})$$
$$= \text{whole number of links.} \left. \right]$$

When the link length increment equals a measurement then the equation = the chain length.

d = a whole number count of links.

a and 360 - a have a whole number count of links.

Fig. 5-5

$$\left[\sqrt{r2 - (\frac{b}{2})^2} = h \right]$$

$$\left[b = \frac{\text{sine}\frac{360}{N_{(1,2)}}}{2} \times r_{(1,2)} \times 2 \right.$$

= chain link length, an applied practice turnkey length,

$$\left[\frac{\frac{b}{2}}{\text{sine}\frac{360}{N}} = r \right] \qquad \left[\frac{360}{N} \text{ degrees} \right]$$

Fig. 5-6

$r_1 = 1$ unit   N = sprocket tooth count

$r_2 = 24$ units   b = link length increment.

sprockets ratio = 24 : 1

$$\left[\frac{N_2 \frac{r_2}{r_1}}{360} \times a \right. \times \text{link length} + (d \times 2) + \frac{N_1 (360 - a)}{360} \times \text{link length}$$

$$\left[d \div \text{Cos} \frac{a - 180}{2} = \text{gears centers distance} \right] \qquad = \text{chain length} \left. \right]$$

$$\left[\frac{360}{N} = b \right] \left[\frac{b}{\frac{\text{sine} \frac{360}{N}}{2}} = \frac{a - 180}{2} \right]$$

104

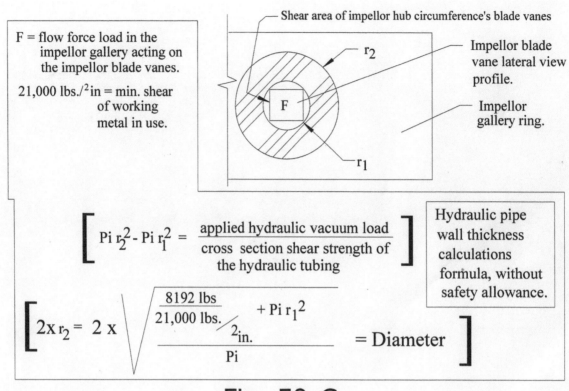

$$\left[\ \text{Pi}\ r_2^2 - \text{Pi}\ r_1^2 = \frac{\text{applied hydraulic vacuum load}}{\text{cross section shear strength of the hydraulic tubing}}\ \right]$$

Hydraulic pipe wall thickness calculations formula, without safety allowance.

$$\left[\ 2x\ r_2 = 2\ x\ \sqrt{\frac{\dfrac{8192\ \text{lbs}}{21,000\ \text{lbs.}}\Big/{}_{\text{in.}}^{2} + \text{Pi}\ r_1^2}{\text{Pi}}}\ = \text{Diameter}\ \right]$$

Fig. 50-C

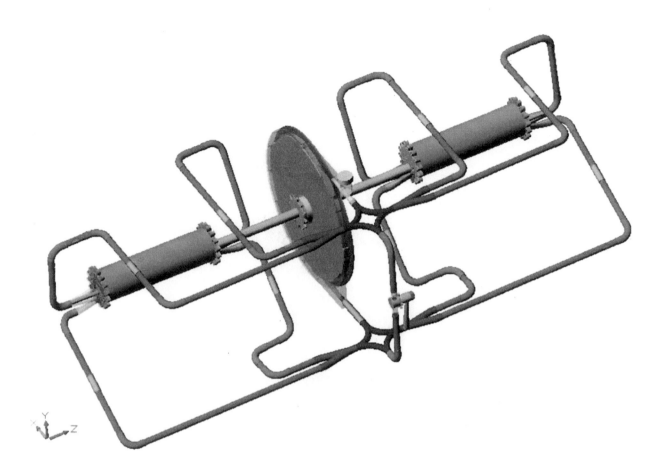

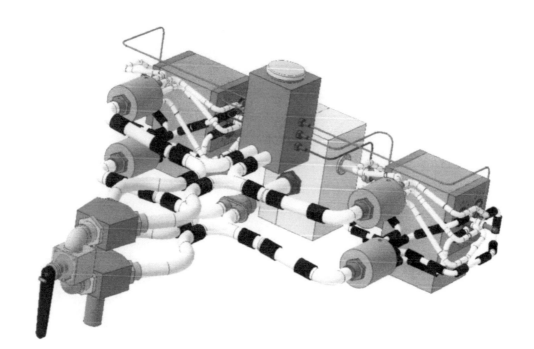

[cubic units]

density of the hydraulic fluid, and volume of the impeller blade vanes circuit of revolution.

$$\text{density} \times \text{volume} \times \left(\frac{\frac{Force}{density \times volume} - 2^n}{2^n} + n \right) \times rpm \times \text{volume of the impeller gallery clearance} \times \text{the fold exponent at the impeller} = A.$$

$$\frac{A}{A - \text{additional clearance volume}} \times \text{the original clearance} = (\text{the new clearance}).$$

see fig. 12-2

The additional clearance volume =

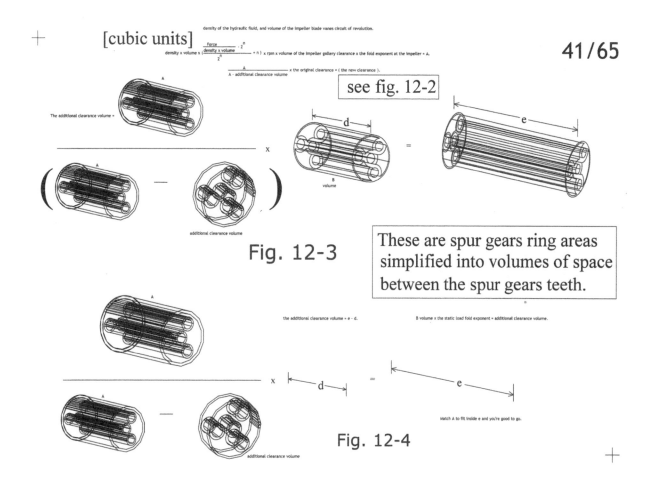

Fig. 12-3

These are spur gears ring areas simplified into volumes of space between the spur gears teeth.

the additional clearance volume = e - d.

B volume x the static load fold exponent = additional clearance volume.

Match A to fit inside e and you're good to go.

Fig. 12-4

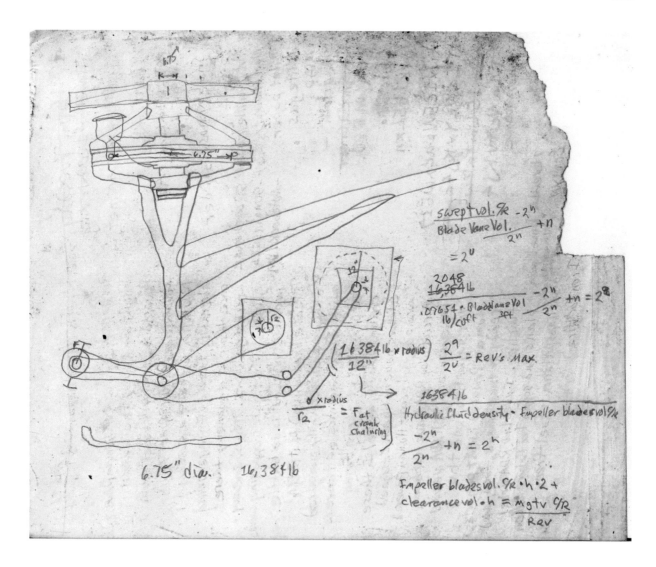

$$\frac{\text{swept vol. }\%_R - 2^h}{\text{Blade Vane Vol.}} \quad \frac{}{2^n} + n$$

$$= 2^u$$

$$\frac{2048}{16,384\,lb} \quad \frac{-2^h}{2^n} + n = 2^g$$
$$\frac{.07651 \cdot \text{Blade Vane Vol}}{lb/cu\,ft \quad 3ft}$$

$$\left(\frac{16384\,lb \times radius}{12''}\right) \quad \frac{2^g}{2^u} = Rev's\ max.$$

$$\frac{\theta \times radius}{r_2} = \text{Fat}\atop \text{crank} \atop \text{Chainring}$$

$$\frac{16384\,lb}{\text{Hydraulic fluid density} \cdot \text{Impeller blades vol }\%_R}$$
$$\frac{-2^h}{2^n} + n = 2^h$$

Impeller blades vol. $\%_R \cdot h \cdot 2$ +
$$\text{clearance vol} \cdot h = \frac{mgtv\ \%_R}{Rev}$$

6.75" dia.   16,384 lb

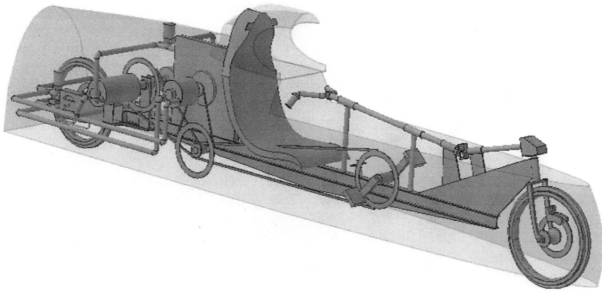

$$\frac{8192 \text{ lb}}{53.9 \text{ lb}/_{3ft} \cdot I_1 bvv} - 2^n \bigg/ 2^n + n = 2^{h_1}$$

99

$$\text{clearance vol}_1 \cdot h_1 + I_1 bvv \cdot h_1 \cdot 2$$
$$= imgtv_1$$

1   18

5

$$\frac{568.89 \text{ lb @ 18" rad.}}{53.9 \text{ lb}/_{3ft} \cdot I_2 bvv} - 2^n \bigg/ 2^n + n = 2^{h_2}$$

$$\text{clearance vol}_2 \cdot h_2 + I_2 bvv \cdot h_2 \cdot 2$$
$$= imgtv_2$$

$$\frac{8192 \cdot \overset{approx}{1.25 \text{ rad}}}{18" \text{ rad.}} \cdot \frac{\overset{approx}{1.25 \text{ rad}}}{5" \text{ rad.}} = \begin{array}{l} \text{operator's} \\ \text{force @} \\ \text{the crank.} \end{array}$$

1.25 = center gear radius.

$$\frac{I_{1/2} bvv \cdot h \cdot 2}{\text{working depth area}} = \begin{array}{l} \text{gear depth both} \\ \text{sides.} \end{array}$$
$$\frac{\text{sum of PTU mgt}}{2}$$

Ibvv = Impeller blade vanes volume
imgtv = initial meshing gear teeth volume

$$\frac{8192\,lb}{53.9\,lb/3ft \cdot I_1 bvv} - \frac{2^h}{2^n} + n = 2^{h_1}$$

clearance $vol_1 \cdot h_1 + I_1 bvv \cdot h_1 \cdot 2$
$= ingtv_1$

199

18
5

$$\frac{568.89\,lb\ @\ 18"\,rad.}{53.9\,lb/3ft \cdot I_2 bvv} - \frac{2^h}{2^n} + n = 2^{h_2}$$

clearance $vol_2 \cdot h_2 + I_2 bvv \cdot h_2 \cdot 2$
$= ingtv_2$

$$\frac{8192 \cdot 1.25\,rad.}{18"\,rad.} \overset{approx}{\cdot} \frac{1.25\,rad.}{5"\,rad.} \overset{approx}{=} \text{operator's force @ the crank.}$$

$1.25 = $ center gear radius.

$$\frac{I_1 bvv \cdot h \cdot 2}{\text{working depth area}} = \text{gear depth both sides.}$$

sum of PTU mgt
$$\frac{}{2}$$

.875

5."4375 −

4.25

$\dfrac{131 \quad 50}{103^{3in}} \quad -1$

$\dfrac{35,185,84\,3in}{703^{3in}} - 2^n/2^n + n = 2^u = 2^{8.33441} \, -1$

$\dfrac{2048}{\frac{.07651 \cdot 1033in}{12^3}} - 2^n/2^n + n = 2^x = 2^{18.71308}$

$449,074.3642 - 2^{18}$

$\dfrac{2^{18.71308}}{2^{7.33441}} = \boxed{11.37867\ Rev}$

$279,231\,73\,lb$

$2048 \xleftarrow{7.76902} 1 \quad v$

$2048 + 279, in = \boxed{2327\ 1/4\ lb}$

$\dfrac{2327.25 \times 1.5\ rad.}{18in} = \boxed{193.92\ lb}$ Primary impeller force

$\dfrac{193.92}{\frac{53.9 \cdot 7.12^{3in}}{12^3}} - 2^u/2^n + n = 2^{9.71}$

$7.12 \cdot 9.71 \cdot 2 \cdot \overset{16}{11.38} Rev$

$= 2212.33\ 3in$

Reduce the impeller diameter.
Raise the force throughout
to raise the operator
force to limits will minimize
all the PTU parameters.
Make a table of impeller radii recalculating each radius for ngtv and operator's force.

Impeller force

Impeller rad.	impeller bvy	Pvalue	3in	Resulting # to secondary PTU	operator's force
13"	10.41	9.55	3,183.1		
12"	5.69	465.45 11.28	2053.9	465.45	174
7"	8.05	332.47 10.293	7651.4 102" 332.47		249 lb.
10"	7.27	349.08 10.35	2006.16		
		13.20			

ngt depth of secondary PTU is 4.26 ft = 51.1495 inches, per PTU(2)

51.42"

24 Rev's.

$$\frac{\frac{193.92}{53.9 \cdot 7.12^{3/in}}}{12^3} - \frac{2"}{2"} + n = 2^{9.71}$$

$$7.12 \cdot 9.71 \cdot 2 \cdot 11.38 \, rev$$
$$= 2212.33 \, {}^{3in}$$

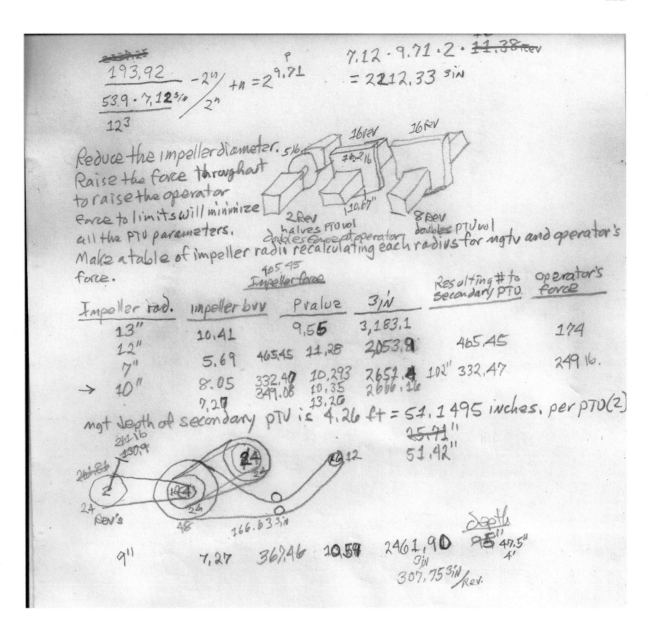

Reduce the impeller diameter.
Raise the force throughout
to raise the operator
force to limits will minimize
all the PTU parameters.
Make a table of impeller radii recalculating each radius for mgtv and operator's
force.

Impeller rad.	impeller brv	465.45 Impeller force P value	3in	Resulting # to secondary PTU	operator's force
13"	10.41	9.55	3,183.1		174
12"	5.69	465.45 11.28	2053.8	465.45	
7"	8.05	332.47 10.293	2657.4 102" 332.47		249 lb.
→ 10"		349.08 10.35	2606.16		
	7.27	13.26			

mgt depth of secondary PTU is 4.26 ft = 51.1495 inches. per PTU(2)
25.71"
51.42"

9"   7.27   36)46 10.57   2461.90   depth
307.75 ³in/Rev.

113

output impeller Vol

1.76715   43,52 rev/sec. $\dfrac{2080\,lb}{\dfrac{53,9\,\cdot\,1.767\,3in}{12^3}}-2^n/2^n+n=2^{15.15}$

$$\dfrac{\dfrac{1.767\cdot 15.15\cdot 2\cdot 43.52\,rev}{16\,rev}}{\dfrac{(1.5^2\cdot\pi-1.25^2\cdot\pi)}{2}}=245.7\;3in=33.7''\,depth$$

$2080\times 1.5 = $
radius
of impeller

18"	173
15"	208
12"	260
10"	312
9"	346.6
6"	520

346.6

$9,5^2\cdot\pi\cdot.25-9^2\cdot\pi\cdot.25=7.26\;3in\;I\,Luu$

$$\dfrac{\dfrac{7.26\,3in\cdot 10,5\cdot 2\cdot 16\,rev}{4}}{(1.5^2\pi-1.25^2\pi)/2}=141''$$

$\begin{array}{l}520\\346.6\end{array}$

$\dfrac{53.9\cdot\dfrac{4.91}{2.77}}{12^3}-2^n/2^n+n=2^{10.5}$

$2^{11.65}$

$$\dfrac{\dfrac{4.91\cdot 11.65\cdot 2\cdot 16}{4}}{\dfrac{(1.5^2\pi-1.25^2\pi)}{2}}=106''\,depth$$

$\dfrac{212''}{6}=35.3''\,depth$

6
mains
PTU to
drive the
secondaries

4ct

4.4'   4.4'   6"rad.

33,7"

Secondaries

$\dfrac{312}{261}\times 6=7.18\;rad.$

at impeller   16 520

346.6

$$9.5^2 \cdot \pi \cdot .25 - 9^2 \cdot \pi \cdot .25 = 7.26^{3in} \, I \, luv$$

$$\frac{7.26^{3in} \cdot 10.5 \cdot 2 \cdot 16 rev}{4} = 141''$$

$$\frac{520}{346.6}$$ $$\frac{(1.5^2 \pi - 1.25^2 \pi)/2}{}$$

$$\frac{346.6}{53.9 \cdot} \quad \frac{4.91}{2.77} \quad -2^n / \quad +n = 2^{10.5}$$

$$12^3 \qquad /2^n \qquad 2^{11.65}$$

$$\frac{4.91 \cdot 11.65 \cdot 2 \cdot 16}{4} = 106'' depth \qquad \frac{212''}{6} = 35.3'' depth$$

$$\frac{(1.5^2 \pi - 1.25^2 \pi)}{2}$$

mains
PTU to
drive the
secondaries

4ct

6" rad.

4.4'   4.4'

33.7''

Secondaries

$$\frac{312}{261} \times 6 = 7.18 \, rad.$$

These six mains
Turn four revolutions per two
cranks at the operator and
deliver the required volume of
hydraulic fluid to turn 16
revolutions at the primary
impeller for the secondary
pumps.

$$7.18^2 \times \pi \cdot .25 - 6.93^2 \cdot \pi \cdot .25$$
$$= 2.77^{3in}$$

$2.77 \cdot 12.46 \cdot 2 \cdot 16 = 1105.28 \ ^3\!/in$

$$\frac{1105.28 \ ^3\!/in}{7.20 \ ^2in} \bigg/ 6 \text{ pTUs}$$

$= 25.36 \ in \ \text{gear depth}.$

7.18 rad.

33,7

25,36

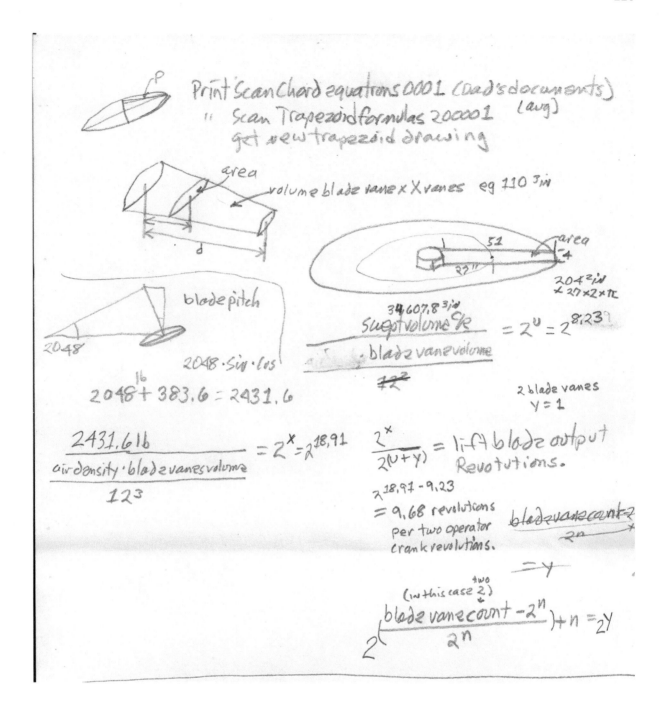

Print Scan Chord equations 0001 (Dad's documents)
" Scan Trapezoid formulas 200001 (avg)
get new trapezoid drawing

area
volume blade vane x X vanes   eg 110 ^{3}in

area

51

$\neq 7$"

204 ^{2}in
× 27 × 2 × π

$$\frac{34,607.8^{3}in}{\text{swept volume } \%} = 2^{U} = 2^{8.23}$$

blade pitch

2048

2048 · sin · cos

$2048^{16} + 383.6 = 2431.6$

2 blade vanes
y = 1

$$\frac{2431.6\,lb}{air\,density \cdot blade\,vanes\,volume} = 2^{x} = 2^{18.91}$$
$$123$$

$$\frac{2^{x}}{2(U+y)} = lift\,blade\,output$$
Revolutions.

$2^{18.91-9.23}$
$= 9.68$ revolutions
per two operator
crank revolutions.

$\frac{blade\,vane\,count-2}{2n} = y$

two
(in this case 2)
$$2^{\left(\frac{blade\,vane\,count - 2^{n}}{2n}\right)+n} = 2y$$

Lift impeller ~~minor~~ diameter = 13.5"
lift impeller depth = .25"
lift impeller blade ~~sides~~ sides clearance depth = .125"
lift impeller blade top clearance depth = .375"
lift impeller major diameter = 14"

hydraulic fluid
density = 53.9 lb/ft

Profile
view

$$(7^2 \pi \cdot .25 - 6.75^2 \pi \cdot .25) = 2.69981 \, ^3in$$

$$\frac{2431.6 \times 27"}{6.75"} = 9726.4 lb$$

$$\frac{\dfrac{9726.4 lb}{\dfrac{53.9 \cdot 2.69981 \, ^3in}{12^3}} - 2^n}{2^n} + n = 2^{16.76236}$$

$$2.69981 \, ^3in \cdot \overset{fold}{16.76236} \cdot \overset{Rev}{9.68} = 438.1 \, ^3in / \dfrac{16 \, Revs}{2 \, PTUs}$$

multiply by 2 here if the result is at half strength.

area x 3 = 6.47953 ^2in

$$\frac{438.1 \, ^3in / \dfrac{16 \, Revs}{2 \, PTUs}}{6.47953 \, ^2in} = 2.11299 \text{ gears depth}$$

add clearance volume:

$$7^2 \times \pi \cdot .125 - 6.75^2 \cdot \pi \cdot .125 = 1.3499 \, ^3in$$

$$7.0625^2 \cdot \pi \cdot .375 - 7^2 \cdot \pi \cdot .375 = 1.03544 \, ^3in$$

sum = 2.38534 ^3in × 16.76236 · 9.68 = 387.1 ^3in
by 2 again

$$438.1 + 387.1$$
$$\underline{825.2 \, ^3in}$$
$$\frac{825.2 \, ^3in}{6.47953 \, ^2in}$$
$$\frac{16 \, Rev's}{2 \, PTUs}$$
$$= 3.98 in \text{ gears}$$

Initial power transmission unit (PTU)
meshing gears depth.

$$\frac{\dfrac{9726.416}{53.9 \cdot 2.69981^{3in}}}{12^3} \;/\; \frac{-2^n}{2^n} \;+\; n = 2^{16.76236}$$

$$2.69981^{3in} \overset{fold}{\cdot} 16.76236 \overset{Rev}{\cdot} 9.68 = 438.1^{3in} / \frac{16^{Revs}}{2^{PTUs}}$$

multiply by 2 here if the result is at half strength.

$$area \times 3 = 6.47953^{2in}$$

$$\frac{438.1^{3in} / \frac{16^{Revs}}{2^{PTUs}}}{6.47953^{2in}} = 2.11299 \text{ gears depth}$$

add clearance volume:

$$7^2 \times \pi \cdot .125 - 6.75^2 \cdot \pi \cdot .125 = 1.3499^{3in}$$

$$7.0625^2 \cdot \pi \cdot .375 - 7^2 \cdot \pi \cdot .375 = 1.03544^{3in}$$

$$\overset{sum}{=} 2.38534^{3in} \times 16.76236 \cdot 9.68 = 387.1^{3in}$$

by 2 again

$$\frac{438.1 + 387.1}{} \\ \frac{825.2^{3in}}{6.47953^{2in}} \\ \frac{16 Rev's}{2 PTUs} \\ = 3.9\cancel{8} \text{ in. gears depth.}$$

Initial power transmission unit (PTU) meshing gears ~~depth~~.

The tail rotor impeller and power transmission are similar.

PTU impeller diameter minor dia. = 33,884

major dia. = 34,382

side clearance = .125

top clearance = .375 × .0625

$$\frac{9726.416 \times 1.5"}{16.942"} = 861.15 \, lb$$

$(34.382 \div 2)^2 \cdot \pi \cdot \overset{.125}{} - (33.884 \div 2)^2 \cdot \pi_{\times 2_{in}}$

$= 3.34 \, ^3in$

$17.254^2 \cdot \pi \cdot .375 - 17.191^2 \cdot \pi \cdot .375$

$= 2.56 \, ^3in$

sum = 5.89 ^3in

$$\frac{\dfrac{861.15 \, lb}{53.9 \cdot 6.675193 \, in}}{12^3} \overset{-2^n}{\Big/_{2^n}} + n = 2^{12.00974}$$

and by 2 again if at half strength

$6.67519 \times 12.00974 \times 16^6 = 1282.68 \, ^3in$

$5.89 \times 12.00974 \times 16 = 1133.1 \, ^3in$

sum = 2415.73 ^3in @ 861.15 lb

$$\frac{861.15}{18} = 47.9 \, lb$$
per each PTU.

$$\frac{2415.733 \, in}{6.47953 \, ^2in} = 5.178 \, in \text{ gears depth}$$

$$\frac{4 \, Rev}{18 \, PTUs}$$

861.15

24 tooth sprocket

861.15 sum of all PTUs

PTU center gear

12 tooth sprocket

Crankarm

258 lb

24 tooth Sprocket

center

$+x$

and by 2 again if at half stength

$6,67519 \times 12.00974 \times 16^6 = 1282.68 ^3in$

$5.89 \times 12.00974 \times 16 = 1133.1 ^3in$

$SUM = 2415.73 ^3in @ 861.15 \, lb$

$$\frac{861.15}{18} = 47.9 \, lb \text{ per each PTU.}$$

$$\frac{2415.73 ^3in}{6.47953 ^2in} = 5.178 in \text{ gears depth}$$

$$\frac{4 \, Rzv}{18 \, pTUs}$$

861.15

24 tooth sprocket

$\rightarrow$ 861.15 sum of all PTUs

PTU center gear

12 tooth sprocket

crankarm $\rightarrow$

258 lb

24 tooth Sprocket

The object will be to make the PTU center gear as small as possible while getting nominal sprocket radius activity to meet operator capability.

$$\frac{2327.25 \cdot 28\frac{1}{2}}{186.\cancel{85}} = \cancel{5102.1} \ 16$$

$$\frac{10,204 \ 16}{53.9 \cdot} \quad \frac{9826.2}{53.9 \cdot 2.69981} = 2^{16.7804}$$
$$\qquad\qquad \frac{}{12^3}$$

$$2.69981 \cdot 16.7804 \cdot 2 \cdot 11.37867 \text{ Rev}$$
$$\frac{1096.75}{1031.3 \text{ in}} = \cancel{80 \text{ in}} \quad 5\text{in gear depth secondary PTU}$$
$$\frac{2.15.3}{2} \qquad\qquad 5.31$$
$$\overline{16 \text{ rev.}} \qquad\qquad 5.5 \ \text{included ptu clearance roughly}$$

chanel clearance at performance:

$$.34438 \cdot 16.7804 \cdot 11.37867 = 65.75 \ ^3\text{in}$$
$$\ _{3\text{in}} \qquad \ _{exp.} \qquad \ _{Rev's}$$

16.942" radius secondary PTU impeller

$+.25 = 17.19172$ radius

$$\frac{869.98 \ 16}{53.9 \cdot 6.69475} = 2^{12.01712}$$
$$\overline{12^3}$$

$$6.69475 \cdot 12.01712 \cdot 2 \cdot 16 = \frac{\overset{2736.85876}{2574.45198} \ ^3\text{in}}{\frac{2.15 \cdot 3}{2}} = \cancel{22}" \ 11"$$
$$\qquad\qquad\qquad\qquad\qquad\qquad \overline{18} \qquad \quad 11.78$$
$$\qquad\qquad\qquad\qquad\qquad\qquad\qquad\qquad 12"$$

channel clearance
$$.84466 \cdot 12.01712 \cdot 16 = 162.90678$$

$$\frac{100.2 \times 1.5}{7.18} = 20.9316$$

$$\frac{20.9316}{53.9 \cdot 5.83} = 2^{6.8} \qquad 5.83 \cdot 6.8 \cdot 2 \cdot 16 = 1268.3$$
$$12^3$$

$$\frac{1268.3^{3in}}{2.15^{2}_{in} \cdot 3 \cdot 2} = \frac{16.4'' \, depth + 4''}{4 \, Rev} = 4.09'' + 4''$$
$$6 \, _{PTUs}$$

$$\frac{416.47}{2.15 \cdot 3 \cdot 2} = \frac{16.15 + 4''}{4 \, Rev} = secondary \, PTU \, gears \, depth$$
$$\frac{2}{76} \qquad = 1.26''$$

210
~~173~~
~~152~~
~~125~~
~~75~~
~~33~~
~~34~~
~~29~~
2
~~161~~
~~144~~
144

1.26''

20.15

9

8''

Perpetual Motion Engine

$$\frac{128\#}{53.9 \cdot 3.19068^{3in}} = 2^{10.25598}$$
$$12^3$$

$$3.19068 \cdot 10.25598 \cdot 2 \cdot 1$$
$$= \frac{65.44708^{3in}}{2.15^{2}_{in} \cdot 3} = 5.07342 in \, gears \, depth$$
$$\frac{}{2} \qquad 7''$$
$$add$$

$$\frac{3in}{72.04311} \cdot .40574 in \cdot 10.25598 = 4.16124^{3in}$$
$$-2054 \, 3in! \, error$$

$$\frac{69.60832^{3in}}{2.15^{2}_{in} \cdot 3} = 5.89599$$
$$\frac{}{2} \qquad ~~11.17~~ \, in \, deep$$
$$5.5\#$$

$$10.25596$$

2.35252^{3in}

X

$\times P$

Now add PTU clearance at fold exponent of applied load

.875
5".1375 –

4.25

$\dfrac{131 \quad 50}{-1 \quad 103\,3\text{in}}$

$\dfrac{35,185.84\,3\text{in}}{103\,3\text{in}} - \dfrac{2^n}{2^n} + n = 2^u = 2^{8.33441}_{-1}$

$\dfrac{\dfrac{2048}{.07651 \cdot 103\,3\text{in}}}{12^3} - \dfrac{2^n}{2^n} + n = 2^x = 2^{18.71308}$ $449,074.3642 - 2^{18}$

$\dfrac{2^{18.71308}}{2^{7.33441}} = \boxed{11.37867 \text{ Rev}}$

$279,231.73\,lb$

$2048 \swarrow^{7.769048}_V \dfrac{1}{1}$ $2048 + 279,\ldots = \boxed{2327\,1/4\,lb}$

$\dfrac{2327.25}{53.9 \cdot 10.41} - \dfrac{2^n}{2^n} + n =$

$\dfrac{2327.25 \times 1.5\,rad.}{18\,in} = \boxed{193.92\,lb}$ Primary impeller force

$2^{12.7498}$

$\dfrac{3019.5\,3\text{in}}{16\,rev}$

$\dfrac{193.92}{\dfrac{53.9 \cdot 7.12\,3\text{in}}{12^3}} - \dfrac{2^n}{2^n} + n = 2^{9.71}$ $7.12 \cdot 9.71 \cdot 2 \cdot \dfrac{16}{11.38\,Rev}$ (ngt area)

$= 2212.33\,3\text{in}$ $= 43.7"$ gear depth

Reduce the impeller diameter.
Raise the force throughout
to raise the operator
force to limits will minimize
all the PTU parameters.

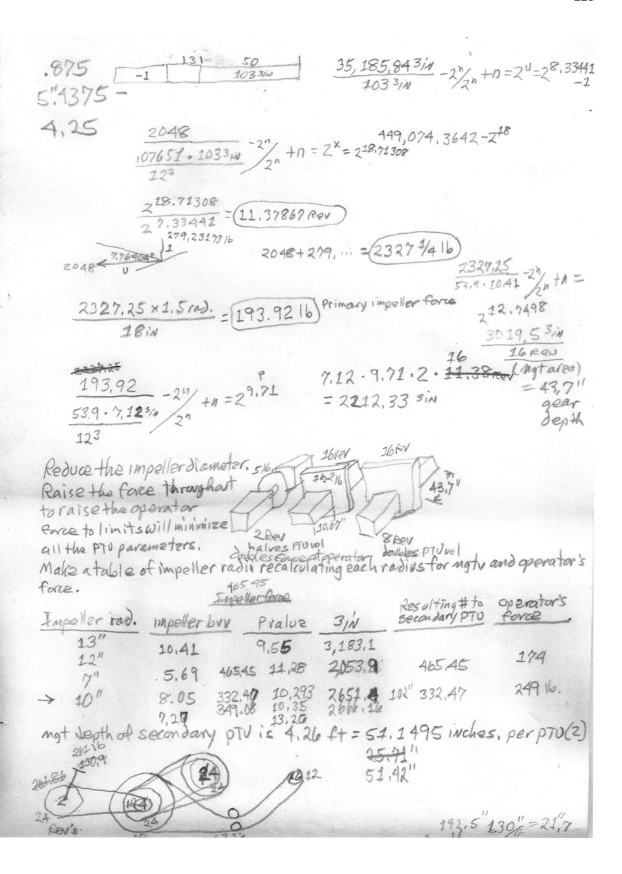

Make a table of impeller radii recalculating each radius for ngtv and operator's force.

Impeller force
465.45

Impeller rad.	impeller bvv	Pvalue	3in	Resulting # to secondary PTU	operator's force
13"	10.41	9.55	3,183.1		
12"	5.69	465.45 11.28	2,053.9	465.45	174
7"					
→ 10"	8.05	332.47 10.293	2657.4	102" 332.47	249 lb.
		349.08 10.35	2806.16		
	7.27	13.26			

ngt depth of secondary PTU is 4.26 ft = 51.1495 inches, per PTU(2)

$\dfrac{25.71"}{51.42"}$

261 lb
130.9

193.5" 1.30"/6 = 21.7"

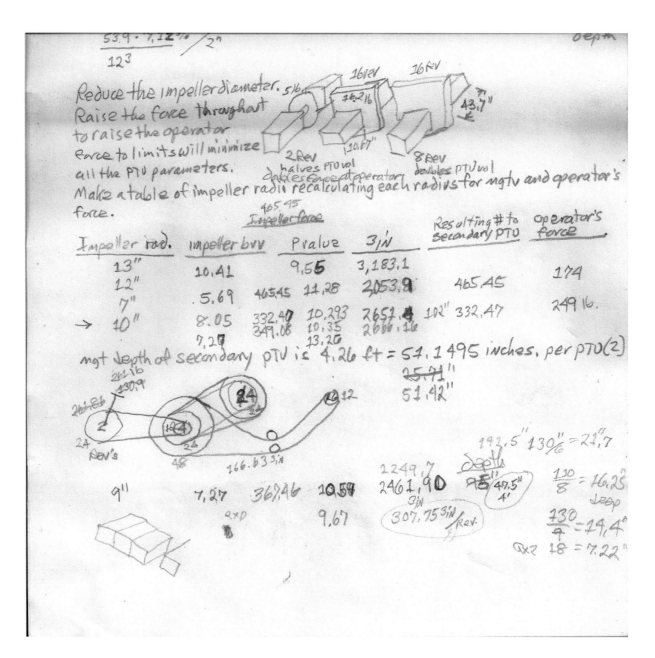

Reduce the impeller diameter.
Raise the force throughout
to raise the operator
force to limits will minimize
all the PTU parameters.
Make a table of impeller radii recalculating each radius for mgtv and operator's
force.

Impeller force

Impeller rad.	impeller brv	P value	3/N	Resulting # to secondary PTU	operator's force
13"	10.41	9.55	3,183.1		174
12"	5.69	465.45 11.28	2053.9	465.45	
7"	8.05	332.47 10.293	2651.4	102" 332.47	249 lb.
10"	7.27	349.08 10.35	2666.16		
		13.26			

mgt depth of secondary PTU is 4.26 ft = 51.1495 inches, per PTU(2)

51.42"

9" 7.27 367.46 10.54 2461.90

307.75 3/N /Rev.

9.67

output impeller Vol

1.76715 43,52 rev/sec. $\dfrac{2080\,lb}{\dfrac{53.9 \cdot 1.767^{3in}}{12^3}} - \dfrac{2^n}{2^n} + n = 2^{15,15}$

$$\dfrac{\dfrac{1,767 \cdot 15.15 \cdot 2 \cdot 43.52\,\text{rev}}{16\,rev}}{\dfrac{(1.5^2 \cdot \pi - 1.25^2 \cdot \pi)}{2}} = 245.7^{3in} = 33.7'' \text{ depth}$$

$\dfrac{2080 \times 1.5}{\text{radius}}$ of impeller $=$

18"	173
15"	208
12"	260
10"	312
9"	346.6
6"	520

346.6

$9.5^2 \cdot \pi \cdot .25 - 9^2 \cdot \pi \cdot .25 = 7.26^{3in}\ \text{I luv}$

$$\dfrac{\dfrac{7.26^{3in} \cdot 10.5 \cdot 2 \cdot 16\,rev}{4}}{(1.5^2 \pi - 1.25^2 \cdot \pi)/2} = 141''$$

$\dfrac{520}{346.6}$

$\dfrac{4.91}{\dfrac{53.9 \cdot 2.57}{12^3}} - \dfrac{2^n}{2^n} + n = 2^{10.5}$ $2^{11.65}$

$$\dfrac{\dfrac{4.91 \cdot 11.65 \cdot 2 \cdot 16}{4}}{\dfrac{(1.5^2 \pi - 1.25^2 \pi)}{2}} = 106'' \text{ depth}$$

$\dfrac{212''}{6} = 35.3'' \text{ depth}$

mains
PTO to
drive the
secondaries

$\dfrac{312}{261} \times 6 = 7.18\text{ rad.}$

4ct   6" rad.

4.4'   4.4'   33.7"

346.6

$$9.5^2 \cdot \pi \cdot .25 - 9^2 \cdot \pi \cdot .25 = 7.26 \; ^{3}in \; I\,lvv$$

$$\frac{7.26^{3}in \cdot 10.5 \cdot 2 \cdot 16\,rev}{4} = 141''$$
$$\frac{}{(1.5^2 \pi - 1.25^2 \pi)/2}$$

$\frac{520}{346.6}$

$\frac{53.9 \cdot \frac{4.97}{2.77}}{12^3}$ $\frac{-2^n}{2^n}$ $+ n = 2^{10.5}$ $2^{11.65}$

$$\frac{\dfrac{4.97 \cdot 11.65 \cdot 2 \cdot 16}{4}}{\dfrac{(1.5^2 \pi - 1.25^2 \pi)}{2}} = 106'' \, depth$$

$\dfrac{212''}{6} = 35.3'' \, depth$

6 mains
PTV to
Drive the
secondaries

$\dfrac{313}{261} \times 6 = 7.18 \; rad.$

4ct

6" rad.

33.7"

4.4'   4.4'

Secondaries

These six mains
Turn four revolutions per two
cranks at the operator and
deliver the required volume of
hydraulic fluid to turn 16
revolutions at the primary
impeller for the secondary
pumps.

$$7.18^2 \times \pi \cdot .25 - 6.93^2 \cdot \pi \cdot .25$$
$$= 2.77 \; ^{3}in$$

$2.77 \cdot 12.46 \cdot 2 \cdot 16 = 1105.28 \;^{3/in}$

$$\frac{1105.28 \;^{3/in}}{7.28 \;^{2/in}} \Big/ 6 \text{ pTUs}$$

$= 25.36$ in gear depth.

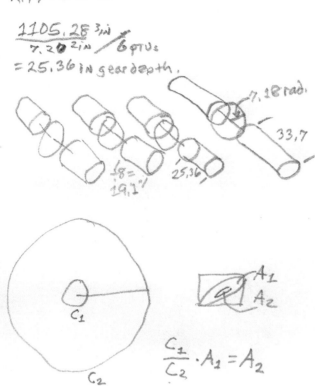

7.18 rad.

33.7

$\sqrt{8} = 19.1"$

25.36

C_1

C_2

A_1
A_2

$\dfrac{C_1}{C_2} \cdot A_1 = A_2$

actual gears diameters: pinion 2.8", 2.937" #8 cutter
gear 4.15", 3.25" #6½ cutter
spaced 3.373"

$$\frac{400^{lb} \times 13''}{5''} \times 2 = 2080 \, lb$$

$$2^{\left(\frac{2080 \, lb}{\frac{53.9 \times ((5.5 \div 2)^2 \cdot \pi \cdot .25 - (5 \div 2)^2 \cdot \pi \times .25))}{123^{3 in/3ft}}} \cdot \frac{-2^n}{2^n}\right) + n} = 2^{76.04888}$$

$$200 \, mi/hr \times \frac{44 \, ft/sec}{30 \, mi/hr} = 293.\overline{3} \, ft/sec$$

$$\frac{293.3 \, ft/sec \cdot 12^{in}/ft}{26 \, in \cdot \pi} = 135 \, Rev/sec$$

$$1.03084 \cdot 76.04888 \cdot 135 \times 2 = \frac{4466.87274}{\frac{(1.5^2 \times \pi - 1.25^2 \cdot \pi) \, 3}{\frac{16}{2}}} = 21.6'' \text{ gears depth}$$

clearance:

$$((5.5 \div 2)^2 \cdot \pi \cdot \tfrac{1}{32} - (5 \div 2)^2 \cdot \pi \cdot \tfrac{1}{32}) \, 2 + (5.5625 \div 2)^2 \cdot \pi \cdot .3225$$

$$- (5.5 \div 2)^2 \cdot \pi \cdot .3225 = .42741^{3in}$$

$$.427413 \, in \cdot 76.04888 \cdot 135 = 926.01602^{3in}$$

$$4466.87274 + 962.01602 = \frac{5392.82826^{3in}}{\frac{(1.5^2 \cdot \pi - 1.25^2 \cdot \pi) \, 3}{\frac{16}{2}}} = 26 \, in \text{ gears depth.}$$

$$26.00897$$

$$\frac{2080^{lb} \times 1.5}{9} = 154 \, lb$$

$$\frac{154 \, lb}{} \quad \frac{-2^n}{} \quad \cdots \quad 9.32795$$

$$\frac{(1.5^2 \times \pi - 1.25^2 \cdot \pi)3}{\frac{16}{2}} \quad \text{gears depth}$$

clearance:

$((5.5 \div 2)^2 \cdot \pi \cdot \frac{1}{32} - (5 \div 2)^2 \cdot \pi \cdot \frac{2}{32})2 + (5.5625 \div 2)^2 \cdot \pi \cdot .3125$

$-(5.5 \div 2)^2 \cdot \pi \cdot .3125 = .42741^{3in}$

$.42741 \, 3in \cdot 76.04888 \cdot 235 = 926.01602^{3in}$

$4466.81214 + 962.01602 = \dfrac{5392.82816^{3in}}{\dfrac{(1.5^2 \cdot \pi - 1.25^2 \cdot \pi)3}{\frac{16}{2}}}$ $= 26 \, in$ gears depth.

26.00897

$\dfrac{2080^{16} \times 1.5}{9} = 1541b$

$2\sqrt{\dfrac{1541b}{53.9 \cdot (9.5^2 \cdot \pi \cdot .25 - 9^2 \cdot \pi \cdot .25)}} \underset{12^3 \, 3in/3ft}{} \quad -2^n \Big/ 2^n) + n = 2^{9.32795}$

$(+2 \,(\text{see below}))$

$\underset{3in}{7.26493} \times \underset{2\pi \, Rav}{9.32795 \cdot 16 \cdot 2} = 2168.54282^{3in}$

clearance 8 $\quad 9.531^2 \cdot \pi \cdot .3125 - 9.5^2 \times \pi \cdot .3125 +$

$9.5^2 \cdot \pi \cdot .0625 - 9^2 \cdot \pi \cdot .0625 = 2.4001^{3in}$

$2.4001 \times 9.32795 \times 16 \times 2 = 716.41781 \, 3in + 2168.54282^{3in}$

$$= \frac{\dfrac{2884.95963 \text{ in}^3}{6.47953 \frac{\text{gears area}}{\text{mat}}}}{\dfrac{8 \text{ PTUs}}{4 \text{ Rev/crank}}} = 73.91382 \text{ in gears depth.}$$

$$\frac{\dfrac{\dfrac{2080 \times 1.5}{9} \times 1.5}{2.5}}{2} = 260 \text{ lb at the crank}$$

operator force $\overset{\times}{\div}$ resulting operator's force

$\overset{\times}{\div}$ diameter of primary PTU impeller yields desired operator's force.

If operator's force is too great:

$\dfrac{\text{diameter}}{\text{resulting force}} \Big/ \text{operator's force} = $ New diameter resulting in the desired operator's force.

all primary PTU's have to be recalculated for gears depth.

$$2^{\frac{716.41781}{3.63247} - 2^n/2^n} + n = 2^{7.54083}$$

$$\frac{53.9 \cdot 716.41781}{12^3} = 22.3466 \times 7.54083 = 168.51189 \text{ lb}$$

$168.51789 \times 9 \times \dfrac{5}{} = 2,022$ lb additional applied force of operator force $\div$ resulting operator's force

$\div$ diameter of primary PTU impeller yields desired operator's force.

If operator's force is too great:

$$\frac{\text{diameter}}{\text{operator's force}} \bigg/ \text{resulting force} = \text{New diameter resulting in the desired operator's force.}$$

all primary PTU's have to be recalculated for gears depth.

$$2 \frac{716.41781}{3.63247} - \frac{2^n}{2^n} + n = 2^{7.54083}$$

$$\frac{53.9 \cdot 716.41781}{72^3} = 22.3466 \times 7.54083 = 168.51189 \text{ lb}$$

$$\frac{168.57789 \times 9}{1.5} \times \frac{5}{2.5} = 2,022 \text{ lb additional applied force of}$$

clearance flow volume throughout the return chain link up from the secondary PTUs.

168.5

2,022 lb

Center   left   Right
5'   5.030 7.607

~~16.085~~ 17.280 19.857
17.134

~~Domain Renewal Notice~~ Now.   Perpetual Motion Engine:

$$9.5^2 \cdot \pi \cdot .25 - 9^2 \cdot \pi \cdot .25 = 7.27^{3}in$$

$$7.27 \cdot 2 \cdot 128 + 2.4001 \cdot 128 = 2,167.036253^{3}in$$

Clearance:

$$9.5^2 \cdot \pi \cdot .0625 - 9^2 \cdot \pi \cdot .0625 + ((9.5 + 1/32)^2 \cdot \pi \cdot .3125 - 9.5^2 \cdot \pi \cdot .3125) = 2.4001$$

$$2.4001 \cdot$$

$$\frac{2,167.036253^{in}}{(1.5^2 \cdot \pi - 1.25^2 \cdot \pi)3 \cdot 2^{2in}} = 167.23 \, in = 14 ft. \quad Waaaay, too \, big!$$
C

$$\frac{541.75891}{c} = 41.8 \, inches = 3.5 \, ft \quad still \, too \, big.$$

$$\frac{135.43973}{c} = 10.45 \, inches = reasonable$$

...and the evening and the morning were The Fifth Day.
Richard L. Chastain © Copr. 2015

Index

Chapter 1

If a bird doesn't have anything to base its design on except an equation with units that have no material concepts, how does the equation evolve? Birds were created long before there were pounds per cubic foot, and cubic inches, and pounds, so a cell has to have something to base its judgment on to develop in to a bird, better still a gene has to have some form or reference to manage the equation's variables values, neither does a bird reference weights or measures and its flight characteristics are inherent and sub-conscious. How do its genes "know" how to develop in to the flight characteristics of the formula for Fluid Mechanics of Inertia without knowing any units, sensing only its environment, as well as into the geometric shape of any aerodynamic structure that flies? This would lead to a construct of trial and error where the successful would multiply and the unsuccessful would perish or develop in to some other form of life, penguins for example.

Nevertheless, not having any units and using successes to achieve the goal eventually the precision of evolution narrows down to minute changes. Still, from the beginning, the prospect of having no units and developing in to a fledgling is puzzling. There must have been much genetic values chosen in combinations before a reasonable solution was made. Still, the values combinations would have been erratic as the genes struggled to align the values for the variables that aligned the bird to the variety of flight characteristics there are in the equation. There are so many kinds of birds. Their values for their variables are all different. A specific species of bird has a particular set of values for the variables. Only as the size and mass of the one bird change the values for the variables the variations are only slight. A different species of bird has an entirely different set of values for the variables yet the equation still solves, the values are significant but not radical. Still, trial and error produced many different kinds of birds. This is a supposition which may later be shown to be a choice made, and different species of birds are derived independently of one another.

It would be possible to use a computer to place only values in the equation for Fluid Mechanics of Inertia, perhaps at random at first but then random selection would reveal close fitting near-answers and refining those near-answers, finding several, and perhaps designing a mathematical bird gene so-to-speak. After several acceptable solutions several designs could be made. Also, the near-answers could be rationalized in to some form of creature. It may even be possible to discover an actual living bird with the solutions. Big did not survive.

Still, having a mathematical solution does not solve the problem of what a gene is able to resolve when it comes to knowing how to be a bird. Genes are adaptive and change with the environment and evolve with reproduction of the species, so the values for the variables also evolve over time. Genes having units upon which to base their judgment would hinder their ability to adapt. Man, however, needs units to shorten the work time of equations. Birds have had time since the beginning of creation to work out the solutions to their equation, and have the rest of time to continue doing the same.

Another evolutionary adaptation was that the genes found that a three dimensional geometric shape being bigger on the top and in the front having its mass center into the larger volume and then to the smaller side of volume opposite the bigger volume of volumes perpendicular was aerodynamically stable both in level flight and on the glide slope. This aerodynamic geometry alignment of values applied with respect to the random assessment of values for variables in the Fluid Mechanics of Inertia equation which, when applied determines the volume and weight of a bird and its air speed at any specific air density. (Air speed here is determined to be more pertinent with respect to individual birds and their respective species and their respective displacement volumes simultaneously in equal time). Different birds of the same species have similar air speeds at equal altitude, while birds of different species have different air speeds from other species of birds at the same altitude. Birds of different species may fly at the same altitude but each individual species has a different air speed for the same air density.

All these factors interact with respect to the bird's consciousness. The bird is consciously aware of its inertia, its displacement, and can feel the force of the air as it flies. Early birds may not have been able to take off and fly but may have started out as glide slope gliders or just flying just above the ground just to gain the feel of flying as would a child learning to ride a bicycle or a fledgling testing its wings. The early Parrot may have been a variety of Pterodactyl, and reasoning this thus-far assuming it to be so then going back even farther in time the Pterodactyl began as a mist of sorts at the beginning of creation, beginning very small, multiplying, gradually learning to fly and becoming a sub-species of Parrot. The Pterodactyl, however, may have been the last of the unsuccessful parrots, like the Neanderthal, that perished in the Nova during the time of Job and killed all the dinosaurs. The smaller animals were able to escape the Nova by finding caves and small shelters to get in to so the Fire from Heaven would not rain down on them and kill them. Many of the comparatively small creatures escaped the Nova.

What other sort of code could a chromosome use for reference for a value for association of the variables in the formula for Fluid Mechanics of Inertia? A gene must associate some means of challenging the equation, and surely there is some kind of chemical or electrical rationale that goes on inside a gene or chromosome that associates the variables for what makes a bird fly (the formula for Fluid Mechanics of Inertia): a chemical for inertia, a chemical for air density, a chemical for body volume displacement, chemicals that tell whether the front is bigger than the back and whether the top is bigger than the bottom and that sense gravity and a chemical that can tell where to put the weight center and make the bird aerodynamically stable and a series of chemicals to tell how to put the inertia where it is supposed to go, and as such perhaps an electrical component. Surely there is something is a birds' genetics that controls all these values for all of these variables and how do they associate with one another to get a solution to the equation? How DID genes associate chemicals and electricity with the equation in prehistoric times, even as far back as when the world was made; this view on the beginning is that the Earth was never a seething inferno but has always been habitable. Had the world been a glowing ball of molten rock and a fiery incinerator from Hell it would have been sterile and lifeless to this day. Accumulations of debris in Space came together over time and knitted our world together non-violently. Surrounding our world were firestorms of Hydrogen and Oxygen perhaps ignited by lightning which rained down water in to our gravity on to our planet. Debris continued to accumulate; peacefully coming together, the water and debris of our Earth eventually became big enough for life to take root if it hadn't already. The planet has always been habitable. As the planet grew it eventually heated up in the interior and the water came out from inside the planet and covered the Earth. Land animals were already millions of years evolved although primitive. Mankind was just starting to populate the planet. The continents drifted apart. Volcanism is a relatively new activity on the Earth now that the Earth is big enough to support volcanism. Wherefore the winged creature...

Insects were on this Earth long before birds, unless of course all of the creatures evolved at the same time, which is highly probable, although it is written that they were not all created at the same time, but that is neither here nor there. Insects may not necessarily look too aerodynamic but they too follow the laws of Fluid Mechanics of Inertia in order to fly. Insects don't seem to have much of a glide slope but they still perform aerobatics just the same. There they too have genetics that must be contended with where chemicals and electrical signals formulate equation relationships of values for variables with no units into living creatures. Flying Insects get considerably small, Gnats even fly. The proliferating creatures of the Earth all strive to survive in an ecosystem of macro-ecology that is woven together and can be fragile. Perhaps the extinction of a mosquito will extinguish a moth which will extinguish a cat, and other multitudes of life threads like that from the producers to the apex creature whether it be a bird or a wasp; extinguishing one pesky life form could possibly extinguish many beneficial life forms for which that one pesky life form was a nail in the coffin of the life thread.

Chapter 2

Unfortunately no research has been done on the prospect of identifying how the genetics of a bird can fill the variables of the formula for Fluid Mechanics of Inertia. It may be so far determined that the chemicals or electrical system in a bird's genes have and still do use trial and error, filling in the values for the variables with chemical or electrical constants to revolutionize the best possible bird. As one species of bird is similar in constants another species' constants are therefore similar in constants although each particular species has a different set of values for their variables. Finding out which genes control the shape of the bird will leave a number of genes for the remainder of the constituencies constituting a bird: bones, sinew, organs (all of which have a distinctive shape to control the entire shape of the bird since their constituency makeup must resemble the bird, and all the organs have their specific locations which is also controlled by the genes, and the organs have their particular inertia), flesh, feathers, legs, feet, claws, and the idea of a functioning bird that lives and breathes and multiplies and flies, and seeks shelter and can defend itself as well as care for its young, build a nest, evade predators, remember, evolve, adapt, and communicate with other birds is all controlled by the genes from the bird's conception throughout its life. The bird is also equipped with minimal bodily regeneration in the event of an injury. At minimum all this information is packed in to and processed by the bird's genes.

Whether or not a bird can think does not lead one to believe that birds are especially profound inventors. Birds also have no necessity for rules, judges, schools, fire departments et cetera: however, their necessity extends to Man and Man's ability to act in defense of a bird or birds should the event arise. Birds benefit from having people around but birds have been on the Earth far longer than Man and as is well pronounced:

Birds would have no problem carrying on without Man on the Earth. There is then the event of cataclysm on the Earth and in Space where a bird would not be able to survive without Man should Man leave the planet at the end of time.

"Decoding" is a relatively weak term in terms of genetics and finding out the arithmetical formulas that make genes produce life forms. Even a simple house fly's brain could only be compared to a comparable computer of today being the size of a planet: Earth, performing all the functions that a house fly has to perform. Building a computer the size of the Earth would probably eventually render out the functions of a house fly, considering the house fly's brain is just that, a brain and is not a miniature computer in any sense. Man has not derived any biological computing capability, but by the time building a computer got to be the size of the Earth there may have been some insights in to biological computing capability rendering all the previous computer technology obsolete and beginning the revolution of biological computing technology. Still, if the computer technology of today was used to finish building the "computer as big as the Earth" in order to mimic the performance of a house fly this may put the house fly at a technological disadvantage and the house fly would have to adapt or go extinct. Then the thread of Life" would unravel for all the food chain from the house fly to the apex creatures which are prone to extinction from the demise of the lower life forms.

The "simple" house fly is exacerbating in its complexity. House flies can be quite aggressive, and a computer with the "intellectual" capability of a house fly would be as a computer that can control the world with violence and harshness and destruction having no law upon which to base any of its judgments. The computer would be without law and any means to enforce a law against a computer that can inflict violence on the Earth would have to be made feasible by being able to pull-the-plug on the computer or shut it off. This, however, is a similar scenario to an old movie from the early 1970's about a dinky little couple of computers on opposite sides of the world that were built inside mountains that managed to take control of his and her own ways and means away from Man and threatened Man with destruction of all life on the Earth in collusion with the other computer. These computers were comparatively small compared to the brain of a house fly.

A house fly has the same elementary genetic complex as other life forms that can fly: it has to live and breathe, reproduce, defend oneself provide for itself, find shelter, couldn't care less about its young, and all the good things that every living creature has to have to survive, very basic for the house fly. Still, a house fly flies, and it's genes have to rationalize the fly's values for its variables in the equation for Fluid Mechanics of Inertia and the evolution of the house fly also is from the mist of a lightning strike in most probability and the genes uses of trial and error to find a solution to what makes a fly fly probably has resulted in the most acute present day state-of-the-art house fly. The beginning of creation when everything was a lightning strike in the atmosphere of a primordial world and there were many lightning strikes the likelihood of which all were stipulated to create something different, simultaneously many of the same flux of lightning strikes created the same creature's mist of microscopic matter in the primordial atmosphere which matter would have drifted down on to the primordial world in where there was the means to spontaneously generate in to the simplest basic life forms.

Since there was an ecosystem to manage the engineering of the macrocosm of living creatures had to be regulated so the environment could flourish. Creatures were developing with the environment while the primordial atmosphere diminished in its properties to conceive life. First, in most respects the more environmentally friendly creatures populated the Earth: then, more creatures could develop that were necessary to manage the ecology of the Earth. Then, creatures with more advanced features in their progress came from the remnants of the primordial atmosphere as the last bit of constituency in the atmosphere was used up by lightning to spark the last remaining evidence of primordial creation finally resulting in the construct of Man. Man may have been a constituent of having both sexes and necessarily, as would be considered "taking a rib" as-it-were, divided into man and woman. All of these things are relevant with respect to the probability that there is a creator, which is more than likely, spontaneous generation of life is not feasible, lightning comes from volcanoes pyroclastic flow and static electricity in the atmosphere, not necessarily the lightning that there is today generated by wind blowing through the life giving foliage on the Earth and being attracted to water molecules in the air thus adhering to one another until the repulsive force of the negative ion can no longer keep the water molecules apart, a bolt of lightning occurs, the water vapor congeals in to rain because the negative ions can no longer keep the water vapor apart and rain falls.. The necessity to multiply in the same manner as all the other life forms, with a few exceptions, was a genetic responsibility necessary for procreation (for lack of a better word), mankind could not divide in order to reproduce, which would make Man microscopic from the beginning and which mist having settled in to primordial sludge found it to be paradise, the result of the last lightning strike(s) in the primordial atmosphere using up all the matter there for producing life and evolving over the Eons into the Man that can make sense to today's reader. In the meantime, all the creatures on the Earth have been evolving for millennia and even perhaps Eons.

Mankind comes along and has a necessity to have to fly. This necessity is brought about by one Believer's interpretation of the Scriptures of the Holy Bible. This interpretation is that in some distant future not too distant but distant enough for the Revelation of flight to come to the Believer and such that the solution to flight results in a simple equation called The Formula for Fluid Mechanics of Inertia, the Earth's Sunlight will be obstructed by the peaceful transit of interstellar dust between the Earth and the Sun. The Sun also will be pommelled by the space debris transiting the solar system and once covered with the debris (interstellar dust) the light from the Sun will be turned red because the debris was cold and covers the Sun to an undetermined depth and is slowly heating up. It shall reflect off the moon and the moon shall appear red. This occurrence will most probably make some strategy necessary to prevent freezing to death in an atmosphere that may very well become liquid without the heat of the Sun. However, this liquidation of the atmosphere may occur in the stratosphere and it may rain liquid air on the Earth from the sky. Thermal-nuclear weapons will be beneficial in helping keep the atmosphere vaporous and the troposphere more habitable. Dropping the thermal-nuclear devices on the ground wouldn't seem too logical when it's not the ground that's freezing, and "the Earth" includes its atmosphere, so thawing out the atmosphere would seem to be a more likely scenario. There would still be "stars" occurring on the Earth, as it were, but "the Earth" would be the part way up in the sky where the atmosphere is turning in to a liquid. Perhaps detonating a thermal-nuclear device deeper in the atmosphere would warm the air and thaw out the stratosphere at the same time; the most efficient use of the thermal-nuclear device would have to be decided. As is unscrupulously obvious, setting off thermal-nuclear devices on the Earth in remote locations doesn't seem to do that much damage to the surrounding population of Man. The nuclear "fallout" doesn't seem to have much of an effect on everyday life as we know it, if in fact there is any nuclear fallout. Where is it? What has it been doing to the environment? How have we suffered from the nuclear fallout of the tests of nuclear devices before the nuclear test ban treaty? Over 2000 thermal-nuclear devices were tested since 1948 and there doesn't seem to be any detrimental effect on life today for ordinary persons. Using the thermal-nuclear devices constructively wouldn't seem to be much bother for everyday life.

Chapter 3

Mankind may have been a lightning strike many Eons ago but by virtue of the fact that there are so many different people on the Earth it leads me to speculate that there may have been many such lightning strikes and using the Holy Bible as an excuse, Man may have been many or a race called Adam (as thus therefore we would call the woman Eve, as a race), as Cro-Magnon may have been called "David" in the Holy Bible which slew Goliath, the Neanderthals or some such metaphor used by the Holy Bible to instead refer to a race using a name of a person; thus, fifteen generations of Man may be spared with reference to a generation being the length of time it takes for there to be a noticeable evolutionary adaptation (60,000 years). If in fact this is so then 15 generations is 9 million years. Now the prospect of Man evolving from a mist that settled to the ground created by a lightning strike in a primordial atmosphere and there being 60,000 years for there to be a noticeable evolutionary adaptation, it would leave there to be 75,000 evolutionary steps from the creation of Man (the resulting mist of a lightning strike in a primordial atmosphere) to the present day if the existence of Man has only been for 4.5 billion years. Make your own judgment on whether or not 75,000 evolutionary steps is sufficient for there to be a Man on the Earth having been "conceived" as a mist in the times of developmental progress of diminishing the materials in the primordial atmosphere in to life forms. Personally this is not enough evolutionary steps (noticeable adaptations) to develop in to the present day state-of-the-art Man. 75,000 may be a lot with one noticeable evolutionary adaptation every 60,000 year, consider a computer generated model of the rate at which a Man would change with 75,000 noticeable changes, it would take 16.67 hours for a Man to revert back to the mist of the lightning strike in the primordial atmosphere. Evaluating what a noticeable evolutionary change would consist of the Man would have to be evaluated to be divisible by 75,000 making minute changes every second which changes would be consistent with any adaptations necessary to cope with the surrounding environment. However, there is an error in the calculations. Man has not been on the Earth for 4.5 billion years: still, 75,000 divided in to however many years Man has been on the Earth according to "Scientists" will result in a noticeable evolutionary adaptation approximately every ten minutes. This is not consistent with the requirement that there be 60,000 years for there to be a noticeable evolutionary adaptation. If Man has been on the Earth for 75,000 noticeable evolutionary adaptations and there is a noticeable evolutionary adaptation every twelve minutes, then Man will have been on the Earth for 2 quadrillion 365 trillion and 2 billion years, save a few tens of billions of years. 75,000 noticeable evolutionary adaptations may not even be enough for Man to revert back to a primordial mist created by a lightning strike. This evaluation is also dependent on there being a noticeable evolutionary adaptation every 60,000 years but what it is really is not really something that anyone can account for by eye-witness but God. So don't believe anyone who comes along and says there is a noticeable evolutionary adaptation every X number of years because they don't know, they're just making it up. 60,000 years was made up by interpretation of the Scriptures of the Holy Bible by the author: still, 60,000 years would seem a reasonable period.

If there are any calculations to perform it is that if the Earth is 4.54 billion years old (for a reference), and that is represented by a 24 hours period and Man has existed for only the final minute on the 24 hour period and there is a noticeable evolutionary adaptation every 60,000 years, and there have been 75,000 noticeable evolutionary adaptations, then Man is 4.5 billion years old from the mist of the lightning strike in the primordial atmosphere. That's very close in time to the prospected age of the Earth, only 40 million years since the creation of the Earth and the strike(s) of lightning that decided that there would be a "Man" on the Earth; which means that all the other creatures and water and light and fish and birds were all decided within the first 40 million years of the Earth's existence, which does not conform to the amount of time for there to be a noticeable evolutionary adaptation although the living creatures are more adaptive and reproduce more often than Man and therefore their time for a noticeable evolutionary adaptation is less than 60,000 years. It doesn't seem feasible for the slowest reproducing life form other than Man to come about in 40 million years with all the necessary noticeable evolutionary steps it would take to go from mist generated by a lightning strike to a state-of-the-art living creature 4.5 billion years ago: however, that is probable since there have been tens of thousands of noticeable evolutionary steps for lower life forms since 4.54 billion years ago to 4.5

billion years ago (over the past 40 million years). It is feasible however, that for a lower life form to evolve from 4.54 billion years ago 40 million years that it would be subsequently comparable to the decision for there to be a Man on the Earth by the last and final lightning strike(s), the lower life forms would have been state-of-the-art for 4.5 billion years ago. Air pollution and lightning may this day be the building blocks of future life forms. Their inception in to life, the primordial mist, would not be evident for 40 million years from now. Perhaps there will be a new kind of bird, not one produced my reproduction but one decided upon 40 million years ago by today's lightning and atmospheric state that has secretly been evolving somewhere out in the wild in an untamed and untouched landscape. 40 million years from now we will find out.

Chapter 4

These are stipulations in the concept of the egregiousness of the chromosome: there is no "cross-talk" between chromosomes. Chromosomes are entities and do not "talk" to other chromosomes. If each chromosome "knows" what it is supposed to do where do these chromosomes get this "knowledge"? What is knowledge to a chromosome? How many generations from the mist were there when the first ideal of knowledge created the first chromosome? What sort of evolutionary step(s) was/were there from the mist of knowledge to the chromosome? How many generations were there?

The chromosome is a microcosm of information but the information is made up of chemicals and electrical impulses. The encyclopedic description of chromosomes is unremarkable. What associations did the Mist make with what it was going to create? It was not the mist it was the lightning that decided what the mist was going to create. The mist was created, which settled out of the sky on to the Earth, and the bearing of the creation was in the mist. The primordial atmosphere eventually displaced, used up by the lightning creating mists. The mists settled on the Earth together and mutually associated. One mist settled on the Earth would form a compound. The compounds became complex and became basic life on the Earth in the form of what would become a winged creature. Speculation on the complexity of the evolution of the mist in to life forms is too complicated, only speculation and hypothesis with prospects on provable results will result in the possible life form. Time is the governor in this draft. Speculation is the only prospect of endeavor for which a solution may be found. Speculate away!

Darwin is a case to be reckoned with. All life on the Earth does not stem from one primordial source. The prospects of explaining the creation of life on the Earth, particularly by species by species cannot develop in to a flow chart of creatures having one original ancestor. As is obvious to the observer there are as many species as there are origins. A plant has always been a plant, specifically by its respective genus and species, and will always be that plant and no transmutation of change will ever occur for all time, that will make that creation something other than the plant it was when it was created by the mist. So is the same for fish, and birds, and mammals, and Man. Their existence was created in the mist, they have always been that creature, and they will always be that creature forever. There are as many origins of species as there are species. Life was not created a single mist of one genus from which all species evolved. If that was so there would only be one creature on this Earth and it would be capable of doing everything. As it were, there are as many creatures and forms of life on the Earth as may be made possible for the Earth to contain them all. If there was more room on the Earth there would be more species. Mankind is a space hog and is using up all the room there is for other creations to exist. How to get rid of the money? If there was less money to be had Man would be less of a space hog using up all the available land for agriculture and progress destroying the very forests that create the lightning in order to feed more people. Less money would be a progressive inventive step, with less dollars to go around prices go down, the dollar becomes stronger.

Since when does a bird or an insect or a blade of grass care about a dollar or anything of value for that matter? Birds are only concerned with themselves and with their survival. The rule of law to a bird is whatever is right in the sight of the bird at the moment: survival of the fittest.

Chapter 5

Of course one does not have to believe that life was started by lightning strikes in a primordial atmosphere but serious doubt exists that would be any other way for the environment to interact to conceive life on the Earth considering all the possible means there are on the Earth for the prospect of environmental engineering of life to take place, lightning would seem to be predominantly the most obvious, but then there is no way to know. Perhaps there was some mysterious manifestation of energy that no longer exists today that caused the life to come to be. Or that energy exists but is on the other side of the firmament, that impenetrable barrier between life and death and the present and all other time which cannot be penetrated into. And the way back is guarded so that Man cannot go back. It is not possible for Man to devolve he can only evolve and adapt so going back can only be speculated on.

If all the foliage and insects and sea creatures and birds and mammals were all constructed in their primitive forms by 40 million years since the inception of Earth and primitive Man was invented at the 40 million years mark, then there would have been the more advanced primitive life before Man's inception and Man would have been like the other creatures concepts were at their beginning 40 million years before and throughout the 40 million years of the other creatures evolution. Man would be a microscopic life form to the other creatures. Man would have already been engineered to become master of the Earth and all there is in it including getting bigger over the billions of years. Man's state-of-the-art size in the present day is also conducive to the environmental engineering of nature. If Man were generally bigger the Earth would most likely have reached its carrying capacity, and if Man were smaller there would be difficulty with fending off predators, to say the least.

Mankind grew up under the noses of other life forms that were already here and most probably was prey on occasion. Still, Man managed to survive and being a predator dominated the Earth as the most superior life form. Obviously then, his brain is the result of the necessity to have to survive, as are all the brains in all the creatures each one in pursuit of one of its own kind, preferably the opposite sex by natural selection. Some of the manifested concepts didn't survive and others did. Mankind's ability to think is a necessity brought about by the necessity to have to survive and with the dexterity of hands with an opposing thumb is able to make things. Natural selection may also have played its part in Man's size. Now Man is coming upon the necessity that one day in the future Man may have to migrate on a regular basis if the Earth is thrown out of its orbit by any account of space debris in its path. If the orbit of the Earth becomes more eccentric and passes out of the habitable zone by any amount the inhabited parts of the Earth near the equator will become too hot and/or the poles will become too cold. The normal region in which Man is situated will have extreme temperature fluctuations perhaps seeing desert heat and polar cold in one day in the temperate zone. As the Earth's eccentric orbit proceeds through the habitable zone some relief will stand. Every three months or so Man will have to pack up and go because the Earth will soon pass out of the habitable zone and Man will have to move North or South to avoid the uninhabitable regions of the Earth during its passage out of the temperate zone. The necessity to have to fly in order to migrate has come into being. Mankind may become less of a permanent resident in any one place migrating from North to South as the temperate region in the troposphere becomes uninhabitable. Mankind may have to revert back to being a hunter-gatherer if these conditions come about. However, there is hope. Eccentric orbits will balance out over time and the Earth may be restored in its habitable zone but it will be a long time once the Earth goes out of the habitable zone before its orbit re-stabilizes and the Earth's orbit may not return to its original path in the event that what disturbed the Earth in the first place either sped up of slowed down the Earth's velocity: if the Earth's velocity is slowed down its orbit will balance out farther away from the Sun and vice versa. But, considering the inertia of the Earth it's not too likely that its velocity will be changed much if it is changed. Such change would be by catastrophe on the Earth such like billiard balls colliding ever so slightly and in space there is no resistance to change in orbital velocity: however, the sizes of the objects impacting the Earth will vary and most likely no object as big as the Earth will collide with the Earth, that would be a drastic change in velocity and the orbital altitude of the Earth from the Sun would change dramatically depending on how fast the object hitting the Earth is going. Cataclysm is impending according to the Scriptures of the Holy Bible. Remember the Boy Scout Motto: Be Prepared.

Black holes are not just that. An area of space where there are no stars may not be a collapsed star but may be interstellar dust, and if the "black hole" is getting bigger over time it is getting closer. Interstellar dust is visible in a galactic panorama. It blocks the light of the stars as seen in the panorama in the interior of the galaxy. Interstellar dust has passed between the

Earth and the Sun in biblical accounts: Exodus 10, 21 -23, and Luke 23, 44-45. (This is only the author's opinion.)

Interstellar dust passing through the Solar System impacting the Sun and covering the Sun with fresh debris will bring about the Revelation to John: 6, 11-17, and the end of chapter 6 where-after it is quiet on the Earth for the space of about a half an hour, which if a day is 1000 years and a thousand years is a day then about a half an hour is about 20.83 years. I suppose Man is getting back on his feet. How long the darkness will last is not written. Man does survive though and there is the beginning of chapter 7 of the Revelation to John. There are also Biblical stipulations involving who survives and who doesn't and one descendant of a Patriarch in particular becomes extinct at the end of the Revelation to John Chapter 6. When Chapter 7 begins all the descendants of that particular Patriarch's son are all extinct (Esau shall perish, Jacob shall prosper).

THE BEGINNING OF THE END

I hope you enjoyed this manuscript of my concept of creation. I have thought for many years on these subjects in this manuscript and have rationalized them to the best of my ability on pages here for you to read again if you shall so choose to. With all my appreciation I whole-heartedly thank you for reading my manuscript and hope that it has given you some substantial vision of what could have been and what may very well come to pass.

Richard L. Chastain

Richard L. Chastain © Copyright 2014
In The End There Was
How To Get Power Out of A Bicycle
By Richard L. Chastain
A Manuscript
(Unedited By Any Publishing Company)
A Variety Of Individually Written Manuscripts On Such Subjects As A Human Powered Helicopter, A Super-fast Bicycle, The Self-Propelled Motor, And The Social Impact Of Human Powered Flight, Including How To Get Power Out Of A Bicycle.

200 mile and hour bicycle
Introduction

This book is not exclusively for highly educated people in the fields of engineering and science and does not exclude graduates of the elementary schools, although it does help to have an Associate's Degree in Mechanical Engineering Technology to start with. If one is able to add, subtract, multiply, and divide then you can do most of the arithmetic problems in this manuscript. You may learn to do some "algebra" and trigonometry, if you don't already know how, which will have to be agreed upon by the elementary school graduates as to the algebras' rearrangements of variables, also some statics, since elementary school graduates may not be too familiar with proportionality and ratios or equations for mobiles, bridges et cetera, certainly not equations for airplanes.

For example: Crossing the equals sign with a variable makes a denominator a numerator and vice versa, and a plus sign a minus sign, a square a square root, and such like reciprocal-like stuff with respect to crossing the equal sign. Also there is the prospect of + + and - - (minus minus) which are both positive and I forget why I needed this though, it was necessary to eliminate a negative solution where a positive solution was required in some statics equations.

The arithmetic is consistent, so don't worry too much about it being difficult. Only the variables (x, y, n, D, C, p, q, u et cetera) will either change or remain the same, while their identity will remain the same. This may seem confusing at first, but the variable for an identity can be different, the identity just has to be consistent with the variable in the formula. The same variable will most probably be used for a different identity to solve other problems with other equations, so the necessity to have to use the same variable exists because there are only so many letters in the alphabet and there are more than 26 identities. You don't have to memorize any of these variables, but the identities are necessary to work the problems whatever variable you choose to use in the equations. Subscripts are used as well to identify variables. Variables are also used for superscripts. Variables are also used to identify bases with superscript variables.

Identities have "units", which are like American Standard Units: pounds per cubic foot, and cubic inches, and cubic inches per cubic foot, and pounds, and things like that. Units, with respect to their "per" identity is the divisor (/), a forward slash, with a e.g. "pounds" in the numerator, a " / " and cubic foot in the denominator so that the relationship looks like this: pounds / cubic foot, "Pounds per cubic foot". The is the equation using "units" only in the formula for Fluid Mechanics of Inertia in the exponents' denominator quantity and the units are cancelled out: (pounds / cubic foot x cubic inches / cubic inches / cubic foot) = pounds. Notice the "units" cancel because the denominator inverts and multiplies to the numerator: pounds / cubic foot x cubic inches x cubic foot / cubic inches = pounds. Then there is the prospect of the dimensionless number, which is required to resolve the exponent of any base, in this case base 2. Pounds / pounds = a dimensionless number when Pounds, in the numerator, is > pounds in the denominator. > (greater than) is a standard symbol.

What started out as an opportunity to speak in front of the class on everything I know about bicycles has turned pretty much in to everything I ever wanted to know about everything I ever wanted to know anything about. Everything I know about bicycles started as a flopped opportunity to speak to the class, when after 30 years or more it has become engaging mathematics of human powered flight, interplanetary ballistics, space science, and rocketry, and using its basic formula to resolve every possible problem in space in force and motion in base 2 using the fold and the cubit (denominator quantity and coefficient to the base two and its exponent (the fold)) for which there must be a conclusion: the fold is the exponent of the base 2, and the cubit is the entire specific denominator quantity of a fraction (This Association Made With The Base Two Exponent And The Particular Denominator Quantity Is A Fact and should be acknowledged as such in tables). I will attempt to explain the performance of the basic equation of Fluid Mechanics of Inertia and apply it to useful functions of trade in today's world: e.g. a jet engine, a helicopter, an airplane, a bicycle, a rocket motor although a rocket motor has nothing to do with human powered transportation it still figures out when it comes to base 2 mathematics (I may not actually know how a rocket motor works but I have drawn a rocket motor concept which may be feasible), and such like devices as they may come to mind.

I write this manuscript for The People who have been left in the dark about everything that makes airplanes fly and helicopters fly and anything else a man or woman can get into and operate like a ship or a submarine, or a truck or a rocket ship and for the benefit of mankind, and for posterity, for those of us who have been kept ignorant of these things for so long, and to defy those who have kept these things secret from us all our lives, i.e. the aircraft industry, educational institutes of aircraft and spacecraft learning, the military, the space industry, and anyone else who would claim to know these things and for those who publish error and false knowledge telling us "the air travels faster over the top of the wing than it does under the bottom of the wing" (which is knowledge in error) when that is all they teach us, and write things after this manuscript is published and perpetrate themselves to be superior in their judgment rather than humble themselves having not written of these things until now for they knew what we were told by the ones who would have us know nothing and did publish information that they were taught or told to publish without knowing whether it is right or not assuming that it's right and in most cases making it far too difficult for the common person to understand.

So that everyone may benefit from my understanding of base 2 mathematics I will endeavor to exhaust the topic of "Fluid Mechanics Of Inertia – The Science Of Mechanics (the engineering of equations)" to the best of my ability in this writing and since it is basically simple mathematics I suspect that when this manuscript is completed that the books that will be left to be written will be so simple that a child can write them. I hope you enjoy this manuscript.

The only shortcoming in this manuscript is the engineering of compound-complex curved aerodynamic shapes to be used for aircraft. The means of assembling these shapes in to usable working models has not yet been discovered. How to construct a working model by assembling a variety of compound-complex curves in to a solid body shape and be able to calculate its geometric center because the shapes are calculable is not yet determined. Thus there remain all the formidable equations remaining to be calculated to find the center of gravity.

Because these compound-complex curves shapes are calculable their subdivisions calculations should be simplified. Having to solve all these equations longhand with paper and pencil and arithmetic using a computer aided design software program to assist in the drawing of the simple to complicated compound complex curves shapes only facilitates the affect that numerous problems must be solved and proven successful before any interest is taken in the development of computer aided design software capable of solving the problems so we don't have to do any, or much, paperwork. Otherwise the problems get ignored and no progress is made in computer aided design (CAD) software programming, the likes for which this mathematical progress is probably incompatible with binary computer programming anyway and a whole new CAD software program will have to be designed for it.

Chapters are short and to the point. Some excerpts from other writings have been added for additional information.

Additional information can be found at www.fluidmechanicsofinertia.wordpress.com, keyed directly in to your browser.

Chapter 1

The Spacecraft

The moment the patch area of contact between the tire and the pavement makes contact with the pavement and the operator mounts the bicycle and begins to apply force to the pedals mathematics are set into play. There are mechanical action and reaction forces, and mechanics of fluid properties that act and react between the bicycle, operator, the equipment on the bicycle, and the surrounding atmosphere at its particular air density for its specific altitude, not to mention inclines and declines with gravity on slopes, all of which can be calculated statically or without motion while the bicycle remains in motion, and including friction which is minimal on a bicycle.

To calculate the applied force at the patch area you will need to calculate the geometric center of the displacement of the bicycle and its rider, and you will need to calculate the center of gravity; you will need to know the geometric volume of the bicycle and its rider which is usually changing continually due to the wind and flapping parts dangling from the operator and/or the bicycle equipment, and the weight of the bicycle and its rider at its weight center and you will have to calculate the aerodynamic shape's center of gravity which changes continually because it is alive. The formula for evaluating applied forces will come later in this writing. Calculating the geometric center and the center of gravity of the bicycle and its rider requires both geometry and statics. The geometric center and the center of gravity are calculated independently of one another. They are separate entities of calculations all together. Once you have calculated the geometric center you will have calculated the geometric volume of the aerodynamic shape simultaneously.

Calculating the center of gravity requires geometry and statics and once you have calculated the center of gravity you will have to calculate the statics to find the weight center since the actual weight center and the center of gravity of an actual bicycle and its rider will probably not coincide in most respects. This leaves the question "How do I calculate the center of gravity?" since the weight center is different from it. Calculating the center of gravity is shown in the drawings on the up-coming pages, it requires the same technique as calculating the geometric volume of the aerodynamic shape to find the geometric center although this time you do not find a center you simply calculate two volumes that are equal (two times) and apply their intersection of two planes.

These concepts will align vector forces in equal and opposite reaction between the wind acting on the bicycle and its rider and the patch area between the pavement and the tire: however, the bicycle and its rider will not behave adequately under high stress aerodynamic conditions because the weight center does not coincide with the calculated center of gravity. This erroneousness between the weight center and the calculated center of gravity is not an error; the weight center simply must be counterbalanced to align the weight center to the center of gravity. The only problem with that is now that the bicycle and its rider is aerodynamically balanced it will probably fly unless the vector between the center of gravity (now the weight center congruent to the center of gravity (BB')) and the aerodynamic center points anywhere else around the center of gravity besides up and back (in the left side view), and going fast may become a problem. However, when the bicycle and its rider are unbalanced like the vector pointing anywhere else problem, the bicycle and its rider are unstable and in an emergency one can lose control and this is probably caused by the aerodynamics of an unbalanced bicycle and its rider. Balancing aerodynamically to create "ground-hugging" would seem to be the best way to aerodynamically construct the aerodynamic geometry of any aerodynamic shape that is designed to move across the ground. Aerodynamic "ground-hugging" may be constructed in to the aerodynamic shape by aligning the vector (center of gravity, aerodynamic center (left side view)) to be down and back with the center of gravity at the top left, and the geometric center of the aerodynamic shape towards the bottom right. This is sort of a mirror of aerodynamic lift and is aerodynamically stable. The sum of the moments may be found equal to zero at the center of the aerodynamic shape with the applied forces of the patch area and BB' at their respective placed distances from the center of the aerodynamic shape. This balancing at the geometric center of the aerodynamic shape is contrary to flight characteristics the sum of the moments equal to zero is not balanced at the geometric center of the aerodynamic shape for a flying machine, and is for a ground-hugging machine. The vector should be down and to the right from BB' in the left side view, the center of the aerodynamic shape being to the bottom right from BB'. Finding the harmony between the volume of the aerodynamic shape and the weight at the center of gravity and aligning the vector angle to apply the nominal force with the minimum of applied power at the optimum velocity to sustain lift requires some evolution, whether you want the lift to go up or down.

Now there is the act of flying a bicycle which is what this entire manuscript is all about. The science fiction of this prospect is doing the calculations and actually engineering a proper bicycle powerplant that will propel a human powered aircraft in to space, not just the surrounding space, where you have room, or into the air...but space, outer-space where there is no air, to achieve an altitude of more than 90 miles high. However that presents a problem because the human powered spacecraft is only propelled by air (don't get any hair-brained schemes that a human powered spacecraft can be propelled by liquid fuel, hypergolic fuel, or solid fuel rockets) and once the air is gone there is nothing to propel the spacecraft any longer, so it will simply fall back in to the atmosphere (there is a prospect of orbital rate combinations to sustain altitude but I don't think any human powered spacecraft will be able to sustain orbit at this time without fuel propelled assistance, since the speed to achieve orbit is considerable and cannot be sustained in the atmosphere): However, getting that much force out of any human powered powerplant design is still a prospect of discovery.

Anyway, the delivery system of the powerplant turbines must maintain the optimum output performance of applied force at the required limits of tolerances to sustain ascent while running out of air density with increasing altitude. This means that the turbine must continually accelerate as air density decreases while maintaining the aligned proper performance output delivery of the thrust on continuous ascent. The negligibility of this prospect is that once the tolerance of the state of flight has been achieved, then being able to sustain a constant cadence while increasing the output of the turbine with decreasing air density simultaneously only applies the required powerplant to be able to sustain lift of flight from the ground and being able to have a range of cadence through the powerplant's power curve throughout the event, as long as the air density is decreasing the turbine will continue to accelerate. A powerplant capable of doing this job may result in the incapacity of the operator to obtain a cadence at lift on take-off. The following concepts are theoretical: It may be necessary to implement a secondary powerplant calculated to succeed to an altitude where it may be jettisoned only to be taken over by another powerplant which powerplant is calculated to engage at an altitude where it will operate through the range of the operator's cadence and succeed to a higher altitude. This presents the problem of having too many stages with respect to which the operator may not be able to obtain normal lift speed at take-off because of their numbers being too heavy overall. A human powered aircraft capable of reaching greater than 90 miles altitude may have to be launched from a mother-ship to reduce the weight of having too many powerplants to jettison after take-off. Also, jettisoning powerplants will alter the weight distribution of the human powered spacecraft and the wings will have to be designed to move to compensate for the weight changes. Still, the turbine can only move air up to the speed of sound, exceeding the speed of sound with a turbine can only be done if the turbine blade vanes displacement is bigger than the spacecraft. Once the air "breaks" from the turbine vanes then the turbine vanes do not develop any thrust in excess of the speed of sound. However, as air density decreases, the sound barrier becomes easier to break. This then presents the problem of turning a turbine that only pumps air where there is low density air. More turbine vanes are needed to pump the same amount of air as air density decreases with increasing altitude.

This theory is the more practical theory: Slow the spacecraft down by increasing its lift capability, align its flight characteristics to air density at 90 miles altitude for its weight (if in fact there is air at 90 miles altitude), apply the most powerful powerplant that this analytical mathematical function is capable of, and build a spacecraft that is so slow on take-off that it will reach optimum or maximum speed with respect to its design parameters above 90 miles altitude. The trick is to design the spacecraft at 90+ miles high "air" density on the ground. Knowing the "air" density at 90 miles altitude is simple enough (if in fact there is "air" at 90+ miles high altitude), although the author won't go into designing an aircraft capable of reaching 90+ miles altitude here.

Constructing a powerplant that will sustain a normal take-off speed on the ground and fly to 90+ miles altitude may result in a very slow cadence at ground level on take-off which can be compensated for by a bypass system which will allow the operator to crank faster while maintaining the normal take-off thrust which would otherwise be a very slow cadence. The bypass system may be gradually closed with increasing altitude as air density decreases allowing the operator to maintain a constant cadence while the turbine accelerates in to less dense air on ascent.

This eliminates the jettisoning powerplants problem and allows there to be only one powerplant. However, the quantity of turbine vanes will be an issue although if the spacecraft is designed at 90+ miles altitude "air" density then the turbine will be designed at 90+ miles altitude and air density and the problems should solve out as a result. Once the spacecraft is built and flies successfully then there will be no question of putting the first bicycle in space remaining.

Chapter 2
The Equation

Balancing an aircraft and designing a human powered helicopter has turned out to be a 24+ years research project on the part of the author. There will be no inventive steps in the development of the final equation shown here in the following figures. On the following pages are drawings and writing, calculations with solutions but the problems are not in numerical form. There problems are in the science of mechanics and the equations are derived from actual values for variables. The drawings and writing are both freehand and computer aided design drawings and writing. Explained in these drawings and writing is the final development of calculating the engine performance, and thrust of a typical aircraft. Also the vector, which is typical for all aircraft, which is the hypotenuse of the lift and drag is finally solved in its discovery using it for all calculations. The final vector found is simple and elementary. It is the vector between the center of gravity (for which the weight center is counterbalanced coincidental to the center of gravity) and the geometric center of aerodynamic shape (the geometric center of the aircraft body geometry), and is placed at the center of gravity plane parallel to the trajectory side view of the aircraft in most cases, the vector being up and to the right in the left side view and passes through the aerodynamic center. Two formulas describe the thrust. The formulas are different but when the values for the variables are applied the solutions are the same. One formula is basic trigonometry; the other formula is the fluid mechanics of inertia formula multiplied to some standard trigonometric ratios the same as in the other formula. Values for variables are in American Standard Units: cubic inches, pounds per cubic foot, and pounds. When concluding the calculations you will have air density, displacement of the aerodynamic shape in cubic inches, location of the center of gravity, the geometric center of the aerodynamic shape, the final vector for calculations, the weight of the aircraft, thrust force in pounds, engine performance in pounds, and a designed model aircraft. The next five or six pages are the drawings and writing of this prospect. Keep in mind, the calculator exponent function does not give the right answer when calculating a decimal point exponent.

Base formula

Initial patch area torque force ratio

Final vector, rough draft

Final vector and calculations

Final vector, completing calculations

Final vector, rough draft

Final vector, CAD drawing

Ground hugging, rough draft
Chapter 3
Balancing An Airplane

Balancing an airplane is relatively simple once you have calculated the aerodynamic center of the aerodynamic shape. Although compound complex curved aerodynamic shapes make this part of the calculating process very hard it is more simple to make aerodynamic shapes from rectangular boxes, prisms, hemispheres, and pyramids or other elementary geometric shapes which can be strung together in to an aerodynamic shape which is more easily calculated for these problems: bisecting the aerodynamic shape at the geometric center with a plane perpendicular to the direction of travel (the trajectory) and applying a parallel plane forward of the first plane until the volume of the aerodynamic shape ahead of the second plane is equal to the volume trailing the first perpendicular plane, will give you the first line of intersection viewed from the left side view of the aerodynamic shape. Next, a plane made through the aerodynamic center of the aerodynamic shape parallel to the trajectory moves a parallel plane below it until the volume below the parallel plane made second is equal to the volume above the plane made first through the aerodynamic center of the aerodynamic shape. The intersection of each second plane of both first planes is the calculated center of gravity. The resulting vector between the intersection of the two second planes and the intersection of the two first planes is the vector which is the final vector (angle w) used to calculate the thrust knowing the turbine blade vanes volume and their specific resulting final vector (angle v) with respect to which the Sines and Cosines for both the aerodynamic shape of the aircraft body displacement and the turbine blade vanes are multiplied ((coefficient) x sine w x sine v + (coefficient) x cos w x cos v) to find the engine performance. The coefficient is the relationship between the weight of the aircraft (BB' in pounds), the air density at altitude (C in pounds per cubic foot), and the displacement of the turbine blade vanes (D in cubic inches), and cancelling cubic inches and pounds/cubic foot with cubic inches/cubic foot) and the formula for Fluid Mechanics Of Inertia: $CD2^{(BB'/(CD/12^3) - 2^n/2^n+1 - 2^n +n)}$ x sine w sine v + $CD2^{(BB'/(CD/12^3) - 2^n/2^n+1 - 2^n + n)}$ x cos w cos v = engine performance force in pounds. This formula is relatively simple, it requires adding, subtracting, multiplying, and dividing. It is simply a complex fraction in the exponent where BB' is the weight of the aircraft in pounds, C is the air density in pounds per cubic foot, D is the volume of the air craft turbine blade vanes in cubic inches, and all the units cancel and the result is a dimensionless number in the fraction in the exponent. Quantities in parenthesis are worked first, or together as quantities, it is difficult to understand in this example what is above the divisor and what is in line with it. CD2 and + n) are in line. The divisor in BB' / (CD/12^3) is in line with – 2^n. You can see the equation in the previous pages of drawing and writing. The equation is relatively simple. 2^n is just the next whole number lower than what $BB'/(CD/12^3)$ is equal to: e.g. 1589 is the quotient: then, 2^{10} is = 2^n because 2^{11} is 2048 which is > 1589 and the fraction in the exponent of the base two will not solve. n is a whole number integer.

These are the equations and calculations of the previous five pages of drawings and writing. Their application is Universal and applies to every creation of the Almighty that moves through a medium and sustains altitude or glides, operates, or simply has force and motion whether it's in equal and opposite reaction to a medium or to itself in space where its impact reaction force sustains it on its trajectory in harmony with attractive gravitational forces in the Universe, or Galaxy to simplify things. By the way, gravity doesn't exist in space, only in matter. Keep in mind the vectors in compression point away from one another, and vectors in tension point towards one another. This is contrary to what is taught in statics courses in college. The reason for these vectors directions is when (1) in compression when the beam is elastic the forces are on the ends of the beam, the vectors show resistance to the forces and their arrowheads face away from one another. (2) In tension the same is true for elastic beams: when the forces are in tension the beam will hold together and the vector forces arrowheads face one another in between the applied tensile forces at the beams ends. If the opposite is true the beams are destroyed.

Thus, helicopters action and reaction forces are simultaneous and in Sum of the Moments equal to zero. The helicopter does not move during static alignment of forces although it is in continuous performance of motion in reality. The helicopter applied moments limits of tolerances are 100% elastic in Fluid Mechanics of Inertia calculations. There is no flex in the helicopter during static alignment of forces but for calculating motion of flex force(s) simultaneously in performance. The helicopter (as well the rocket and the jet airplane, submarine, ship, et cetera) is fixed in time during calculations (zero time present) although it is still allowed to be alive in its perpetual motion since it is a mechanism that operates in space and time and can be manipulated to perform activities of required motion continually while it is performing its output alignment of static limits of tolerances for which "zero time present" is always in alignment with respect to the helicopter's output power, even though the helicopter is constantly changing its curve of performance throughout its life during its flights. As well, all these performance variables apply to birds.

Applying the altitude air density at which the powerplant will apply its optimum output limit and aligning the lift blade vanes displacement and blade vanes angle elevation attitude to sustain maximum altitude for the optimum weight of the aircraft (helicopter), which is virtually impossible to control (the author will write about that later), the helicopter should fly at its best when parameters of power curve maximum, blade vanes volume, output revolutions at optimum power limits of tolerances, weight of the helicopter and any other limits of tolerances which may enhance the resulting performance of the helicopter will manage to provide for a reasonable helicopter design throughout its powertrain.

Now this is the cop-out: most of the previous rationale is for consumable propellant fuel powered helicopters. A human powered helicopter requires the static elastic balancing limits of tolerances with respect to the sum of the moments equals zero. Finding where the torque was was what the Revelation into this human powered helicopter was all about. A torque lever applied can affect a depth of meshing gears which gears depth can be moderate if not extreme since the torque lever and the gears radius does not change, the torque remains the same. Meshing gear teeth fill with fluid and become a pump and then torque applies and the gears depth makes the torque apply to drive a hydraulic pump of epicyclical gears, the force of torque is constant despite the depth of the meshing gears teeth. A chain drive sprocket drives the center gear in a planetary gears pump. The center gear is 1/3 the ratio of the 1 sprocket radius, or vice versa, and can be a different ratio. As well the hydraulic fluid coming from the planetary gears pump turns an impeller with a radius of, say, 45 inches or so which impeller having its hydraulic fluid applied to its blade vanes in suction applies torque to the center gear of another epicyclical gear pump which hydraulic fluid turns an impeller which turns the lift blade.

As well, another set of planetary gears powered by the operator's chain drive assembly powers another impeller which drives another planetary gears pump which turns an impeller which drives the tail rotor blade.

Aligning the impellers radii so the calculations for applied forces throughout the entire powertrain are within tolerances of the operator's ability to apply force to the pedals, align the applied force at the pedal by varying the impellers radii with respect to the operator's capability to sustain the applied force at the pedals. These calculations of alignment of the impellers radii simply require ratio and proportion to get the optimum impeller's radius on the second try. The desired operator's force divided into the actual operator's force calculated x the impeller radius = the new impeller radius which will apply the desired operator's force. The lift blade should be 2/3 of the desired load while the tail rotor force should be 1/3 of the desired applied operator's force. The applied force at the pedals and to the operator should be low enough so that the operator can maintain flight for a reasonable period of time. Human powered helicopters are custom designed with respect to the buyer's specifications: inseam, height, weight, waistline, shoe size, nominal strength.

The conditions on a human powered helicopter apply the same as a combustible fuel powered helicopter; the lift performance and tail rotor performance must remain at flight performance output power throughout the duration of flight. One cannot "coast" on a human powered helicopter without immediately losing lift performance and tail rotor control.

Rocket Motor Sketch, CAD drawing

Rocket Motor Equations, rough draft

Rocket Motor Equations Completed, rough draft

Rocket Motor Equations, CAD program

Rocket Motor Equations, Cad program

Rocket Motor Equations, CAD program and remaining problem
Chapter 4
The Rocket Motor

The previous pages show a mirrored parts attempt to show a rocket motor in its most basic simplicity and some of the mathematics used in Fluid Mechanics of Inertia to design a rocket motor. None of the variables in the assembly drawing are calculated for precision tolerances. The assembly drawing is simply pieced together from scratch and no parts in their specific tolerances align or match with respect to the equations necessary to make a rocket motor work. The following description will be an attempt to describe the mathematics involved in the assembly drawing of the elementarily simple rocket motor on the previous pages.

A rocket motor's combustion must be aligned in performance at static alignment throughout the powerplant. The turbines in the turbochargers are aligned with static alignment, revolutions, and clearance tolerances so the elementary equation for that prospect of idea is relatively simple and will be described in equations. These equations will be in "Fluid Mechanics of Inertia's" base formula, so all there will be involved will be adding, subtracting, multiplying, and dividing, so a sixth grader should be able to design this attempt to describe a rocket motor's performance using the drawing and the base formula for Fluid Mechanics of Inertia.

Practice with the base formula for Fluid Mechanics of Inertia will be applied for the duration until the time when human powered helicopters becomes the subject in this manuscript. In the meantime the first determination to exercise the base formula for Fluid Mechanics of Inertia will be by using a rocket motor to apply the values for variables in the formula to describe the output performance of this particular rocket motor concept in operation. This rocket motor design would seem to be self perpetuating in concept: however, there are certain parameters inside the rocket nozzle which must be aligned for this rocket motor to work. This rocket motor needs combustible propellant and oxidizer to operate.

The fuel and oxidizer inlet ports at the rocket nozzle have to be a certain area which is proportional to the area of the conduit leading to the turbines. The combustion of the fuel and oxidizer apply force to the fuel and oxidizer inlet areas by applying torque force at the turbines to the fuel and oxidizer. The combustion of the fuel and oxidizer also has a standing force at the conduit area at the rocket nozzle that acts in equal and opposite reaction to the resulting force applied at the turbines to overcome the combustion force at their port areas in the rocket nozzle. The turbines sustain equal and opposite reaction force to the force within the conduit to the turbines at the required revolutions to sustain the force of the fuel and oxidizer at their port areas inside the rocket nozzle. The combustion in the rocket nozzle is sustained and can vary with throttle and all the variables must vary simultaneously to sustain the combustion in the rocket nozzle.

Being able to exceed the static force at the port areas for the fuel and oxidizer to the rocket nozzle allows for the rocket motor to increase combustion and raise the force applied inside the rocket nozzle allowing for acceleration of the equal and opposite reaction performance. What exactly these limits of tolerances are would require testing and research to find the exact match for all of the applying variables: rocket nozzle shape, fuel and oxidizer inlet port areas, conduit area to the turbines, turbines applied force to the fuel and oxidizer with leeway for acceleration forces in the turbines output capability, and simultaneous alignment of all applicable variables in performance being accurate.

Since fuels and oxidizers have various ratios to combine them for complete combustion the associated areas will also align along with their particular turbines specifications to sustain all the required limits of tolerances of complete combustion in the rocket nozzle while allowing for acceleration with respect to the turbine's capability to increase output force to the fuel and oxidizer flow to the rocket nozzle under reaction force of combustion against the inlet port areas. In the assembly drawing on the page of the elementary rocket motor concept the components are exaggerated in most respects.

It may not take as much force as one may expect to sustain combustion in a rocket nozzle. Flooding the rocket nozzle with the fuel and oxidizer and igniting it, or having it ignite if it is a hypergolic fuel and oxidizer combination, and continuing to allow the fuel and oxidizer to flow manually, e.g. with starter motors and a battery turning the turbines, will deliver force to the conduit which drives the turbines when the combustion begins, successfully with an excess of force while the governor will allow the engine to start sustaining the equal and opposite reaction forces managing a running rocket motor while a governor will allow the forces inside the rocket nozzle to accelerate the alignment of all the variables to the desired performance output of the rocket motor concept to idle. This requires more research to get a rocket motor output performance which will apply its difference of actual fuel/oxidizer combustion force minus the applied equal and opposite reaction forces of the applied turbines through the conduit area at the rocket nozzle measured at the turbines equal and opposite reaction forces in sum, and the resulting difference of applied force is the rocket exhaust thrust. The applied formula for Fluid Mechanics of Inertia describes the limits of tolerances that a rocket may possess for its rocket motor exhaust thrust force to sustain a comparable successful test of a rocket on the first try.

The fluid mechanics of aerodynamics are taken in to consideration: Atmosphere reacts on the spacecraft from the moment it takes off. The spacecraft must sustain equal and opposite reaction force against the atmosphere plus acceleration in to the atmosphere while if the spacecraft is occupied the acceleration of the spacecraft cannot kill the occupants and the capability of the rocket motor thrust output must have a range through which the rocket motor operates which will sustain the aerodynamic force of the atmosphere as the atmosphere becomes less dense with altitude and the rocket accelerates simultaneously which acceleration cannot kill the occupants as gravity becomes less with altitude as well, the difference in gravity can be transformed in to acceleration of the spacecraft, the rocket motor must be able to sustain the maximum aerodynamic force (in pounds) on the spacecraft's displacement (in cubic inches) up to the altitude where its velocity and the aerodynamic force are at their maximum, increasing altitude with increasing velocity will only maintain the same aerodynamic force or the aerodynamic force will diminish if the spacecraft does not accelerate in to less dense air fast enough. (see supersonic calculations in the following drawing and diagrams (next).) I do believe this altitude moment is called "Go At Throttle Up": gravity is diminishing, atmospheric air density is dropping faster than the spacecraft is accelerating and the spacecraft can get altitude by going straight up with nominal thrust as aerodynamic force diminishes, the spacecraft accelerates faster, the occupants experience constant steady thrust (equal and opposite reaction forces), and the rocket motor performs within its range of power curve with the capability of sustaining the velocity of the spacecraft in equilibrium with gravity at orbital altitude. Some trigonometry is used during orbital entry to balance the apogee and perigee on the orbit of the spacecraft. This leads to the cubit of the Earth (1 unit cubit) to be useful in ascertaining all the variables of interplanetary space travel. Now that man knows how to orbit the Earth with a spacecraft it is possible to use all the variables values of Earth as 1's (units cubit in the denominator) in their relationship with other planets: orbital altitudes, planetary mass, diameter of the planet, anything that can be determined from calculations that are useful for ascertaining planetary ballistics around Earth can be applied to determine values for extraterrestrial bodies provided one can orbit them. Any extraterrestrial body can be orbited, it depends on one's velocity and altitude mostly in comparison to the 1 unit cubit of the respective variable one is using to calculate an orbital trajectory on approach. There are also intervals on approach, how rapidly one is approaching the celestial body. Intervals are coefficient with the altitude and the base two fold exponent. How many altitudes there are between the orbital tangent at altitude and the distance the spacecraft is from the orbital tangent point are proportional with respect to the time it takes to cross the distance and Earth's 1 unit cubit for the same proportional alignment limits of tolerances with respect to the same variables. Planets can be any size and any density, so a planet that's Earth sized could be more dense (or less dense) so if one entered orbit at the same altitude one would enter orbit at Earth one would have to be going a proportional velocity with respect to the planets densities ratio x the velocity unit cubit of Earth at the same altitude = the extraterrestrial orbital entry velocity, or else the same ratio multiplies to the altitude and the spacecraft enters the orbit at the Earth's coefficient 1 unit cubit velocity, and sustains orbit around the extraterrestrial body. All of Earth's proportional coefficients are equal to 1 until cubit in equations involving interplanetary space travel. Remember the cubit is the denominator quantity and a ratio has a denominator which may be a quantity product or some complex fraction quotient. Using Earth's "cubit" in the denominator and multiplying that denominator to some value of some variable making all the units cancel and the ratio a dimensionless number will result in the coefficient of proportion which can be calculated as a base two exponent which base two and exponent when multiplied to the denominator quantity will equal the original extraterrestrial value. Some consideration for a ratio of intervals can be made but that only complicates the problem.

Fly-by-the-seat-of-your-pants planetary orbital entry evaluation

Theoretical Supersonic Calculations
Chapter 5

A Jet Engine

Once the engine performance for an "airplane" has been calculated knowing the v angle of the turbine blade vanes the combustion force inside the jet engine should be equal to the engine performance force calculated solution. But, the torque from the applied moment of the v angle at the blade vane volume with respect to the lever arm from the v angle to the, in this case, crank journal center multiplied to the engine performance solution should equal the internal combustion force in the combustion chamber of a reciprocating engine. As can be reckoned this combustion force must be sustained continually during the aircraft's flight. This combustion force cannot be sustained by ordinary automobile engines because their connecting rods are not designed to sustain the continuous force greater than the weight of the craft being piloted and sustained at altitude for any length of time. Aircraft internal combustion reciprocating engines have connecting rods specifically designed for sustained engine performance force x the torque ratio = the combustion chamber force. AVGAS therefore would also be specifically designed to burn cleaner than regular or premium automobile gas at the sustained power required to fly an "airplane". A jet engine performance force is approximately equal to the calculated engine performance force although it must include the additional aerodynamic force at the intake turbines and would double the applied jet engine combustion force in the jet engine combustion chamber. The jet engine pulls and pushes simultaneously. Half of the engine performance force pushes while the other half pulls. The calculated solution with respect to the engine performance is the same as 1 unit cubit. Other calculations can be made using other units cubit: the turbine blade vanes volume, the displacement of the aircraft body et cetera. The jet aircraft should be designed at its maximum altitude with its body displacement volume and its weight being conformable to the engine performance at its optimum output power for the particular air density it will be operating in at the simultaneous altitude, which calculations will lead up to the human powered spacecraft which design parameters are described previously. No drawings are supplied for any jet engines at this time.

Chapter 6
The Extreme Bicycle

Here is a design for a bicycle capable of reaching a speed of 200 miles an hour. One problem is the head-tube needs to be strengthened because of the statics balancing putting most of the weight on the front end. The design parameters of the extreme bicycle are the same as for the human powered helicopter only the output force powers the rear wheel instead of a lift blade. All the calculations are made using the base formula for Fluid Mechanics of Inertia. Calculations are made to include impeller blade vanes volume circuit of revolution (1 u.c. (unit cubit)), clearance volume in the impeller gallery, torque, meshing gear teeth volume circuit of revolution of the center gear, the power transmission clearance volume, applied forces, fold exponents, operator's force at the crank, aerodynamic displacement volume of the extreme bicycle's body displacement volume, applied aerodynamic force at 200 miles an hour (which requires the unsolved problem mentioned above in drawings, for impact force of air at the velocity where it is equal to its weight), and the patch area road/tire force which translates back through the power transmission assembly to the operator at the crank force.

The impellers' diameter is aligned to apply the particular force at the operator which can be sustained at 200 miles an hour. Revolutions of the impeller in performance are multiplied in to the equations multiplied to the impeller blade vanes swept volume (1 u.c.) and the clearance volume in the impeller gallery, which clearance volume x revolutions is added to the applied performance of the impeller at applied revolutions, and the sum of the moments equal zero as long as the operator's force is compatible to the operator for sustained 200 miles an hour.

Calculating the impeller's final flow volume of hydraulic fluid and dividing by the sum area of the center gear circuit of revolution times the number of meshing points of gears in the power transmission and dividing by two power transmissions (x/2) will provide the depth of gears in the secondary power transmission.

The primary power transmission calculations are the same except divided by 4, or 6 depending on how many there are. The clearance volume in the power transmission is multiplied by the applied load fold exponent to equal the clearance flow volume. Now the clearance flow volume has to be added to the gears depth to sustain its applied force at 200 miles an hour with no hydraulic fluid "slippage". The next page shows the equation for this calculation of making the clearance flow volume add to the gears depth.

Fluid Clearance Standing Wave Calculations

Fluid Clearance Standing Wave Calculations

Fluid Clearance Standing Wave Calculations

First attempt at 200 mile an hour bicycle (not a failure)

200 mile an hour bicycle design prospect

Once all these equations are satisfied the only remaining alignment of values are the drive train sprockets radii. The sprockets radius on the power transmission for delivering hydraulic fluid to the impeller are three times the radius of the center gear of the epicyclical power transmission. These sprockets are driven by a sprocket of equal radius independent of the power transmissions. On the same axle as the independent sprocket is a sprocket that is ½ the radius of the large sprocket. This small sprocket is driven by a sprocket on the crank that is aligned to be double the radius of that smaller sprocket. Then there is the torque between the crank sprocket and the pedal(s) which finally should provide the operator with a comfortable applied force and comfortable cadence up to 128 revolutions per minute or 2 revolutions per second approximately.

A throttle is provided because if the power train were simply fixed the crank force would be too extreme at the beginning of the run. Hydraulic fluid is allowed to bypass the final output impeller and flow to the primary impeller without flowing through the impeller gallery of the output impeller. This allows for the operator to crank a reasonable cadence while the bicycle is being accelerated from start by closing the throttle. As the throttle is closing more hydraulic fluid is being made to go through the output impeller gallery.

The above paragraphs of chapter 6 are the required points of calculations. Other calculations can be made e.g. the center gears radius of the power transmissions and resulting sprocket radii, the bicycle can be slowed down and recalculated, the bicycle can be designed for children and recalculated et cetera.

Tubing wall thickness equations

Tubing wall thickness equations
One other calculation has to be made, the tubing wall thickness, since there is considerable suction on the hydraulic fluid in the draw tubing. The head tubing not so much head force if any at all since the head flow just pushes hydraulic fluid up to the reservoirs. Once the hydraulic fluid is in the reservoirs the other sides of the meshing gears teeth in the power transmissions suctions the hydraulic fluid through the impeller gallery from the reservoir around the impeller blades to the suction meshing gears around and to the head side of the power transmissions' meshing gears where it is pumped back to the reservoirs.

As well, there are centrifugal and centripetal forces to contend with for a bicycle during cornering. Inclined curves calculations can be relatively complicated. The following drawings show such complications:

Centrifugal and centripetal force equations of bicycles cornering

Starting behind the driver, the first "boxes" are the initial power transmission pumps that pump hydraulic fluid to and from the "box" in the middle in the center on the back in which in the "box" in the middle in the center on the back is an impeller that is to be turned by the hydraulic fluid flow force suction from the first "boxes" behind the driver. The initial power transmission pumps for the super-slipstream bicycle are two quad packs of power transmission units having a gears depth of 13.2 inches.

The paths of least resistance to allow fluid flow to follow around the impeller outer limits (of the second "box") are channels designed in to the impeller on either side of the impeller blade vanes. The force of this hydraulic system is designed on suction and aerodynamics. The hydraulic tubing goes in to and out from the initial pumps (left and right side "boxes" directly behind the driver), on both sides of each pump. The hydraulic fluid lines have bleed valves to purge the air in the lines. A light suction will be applied externally to the suction lines at the bleed valves placed along the suction lines to draw out the air in the suction lines since no reservoir can be placed along the suction lines. The bleed valves will be closed once the air is removed from the suction lines. Hydraulic fluid will be drawn in to the suction lines via the grease seals secondary tubing from the reservoirs through the clearance very slowly at first, a light suction will be applied at the bleed valves to draw out the air in the assembly.

The "cylinders", above and behind the "boxes", have conic sections cut in to them on the inside to funnel the hydraulic fluid in to or out from one line of tubing coming in to the opposite ends of the cylinders from the impeller sides. As the hydraulic tubing reaches the impellers (of which there are two, one is at the rear wheel) the internal diameter of the tubing narrows with respect to the depth of the impeller blade vanes (depth is transverse to the tubing length). This, in effect, creates a velocity change of the hydraulic fluid flow to the impeller blade vanes, which fluid velocity is particular to each impeller. This velocity change is only respective to the upper hydraulic tube of the respective impeller. Two revolutions of the crank with a 24 tooth chainring at the operator develops 16 revolutions at the first impeller and 43 revolutions at the final drive impeller, produced by the hydraulic fluid velocity developed by the first four (quad set of) power transmission units (pumps), then the second set of power transmission pumps drive the rear wheel.

The second pair of power transmission units (left and right rear "boxes") simultaneously deliver hydraulic fluid to and from the second impeller located at the rear wheel. The second pair of power transmission units (secondary power transmission units) have a gears depth of 26 inches.

The [view of the] throttle [is obstructed, it] is [behind a tubing in the view,] next to the rear wheel impeller "box" between the two lines of inflow and outflow tubing. The same principles apply to the hydraulic fluid flow from the second set of power transmission pumps ("boxes") to and from the second impeller only the second impellor is a smaller diameter and the pumps deliver more than 8X the fluid flow volume of the initial pumps with two revolutions of the chainring at the pedals (see formulas) due to the fact that the initial pumps only turn four revolutions per two cranks while the secondary pumps turn 16 revolutions per two cranks of the chainring at the pedals. The result is 43 revolutions per second at the final drive impeller per two crank revolutions of the chainring at the pedals.

A 26 inch diameter wheel turning 43 revolutions per second and a 24 tooth chainring with two revolutions of the chainring in equal time at the crank should require the bicycle to just reach a speed of 200 mph. The wind resistance is estimated to be over 400 pounds, it has not been calculated exactly. Software for Fluid Mechanics Of Inertia is required to calculate the exact air flow force to the vehicle.

Certain measures (tubing wall thickness calculations) have been taken to make sure the tubing does not collapse under the suction load, whether these calculations are correct is as yet untried. As of this moment the only difficulty foreseen is the mangling of the suction tubing under load since the suction tubing will probably all try to become straight as load is applied (this will not happen to the head side of the hydraulic tubing since it suffers no load, only to the suction side). These reservoirs have secondary tubing which feeds hydraulic fluid in to where there is a grease seal as a static seal at each central axis of power transmission unit for each "box" assembly and supplies hydraulic fluid (or anti-freeze and water 50/50 ratio, or sewing machine oil) to prevent the grease seals from sucking air in through the grease seals contact area since there will be a great amount of suction in the gears galleries and impellers housings. The suction force should be minimized because of close tolerances. Hydraulic fluid is supplied to these regions by the reservoirs in both locations which reservoirs' hydraulic fluid is at ambient temperature and pressure and have an opening to prevent draw suction. Once all the air is purged from the assembly the applied force to the hydraulic system should be solid. Bleed valves are placed in strategic locations and a light suction at the bleed valves will disperse trapped air from the main hydraulic tubing. The tubing and power transmission unit casings and plating can be made of polyvinyl-chloride or some other kind of light elastic plastic and their components glued together. Complications of tubing assembly can be remedied with a sleeve and a strategic cut made where the tubing is straight and the sleeve is inserted to cover the cut and the assembly succeeded to by gluing the sleeve to the tubing over the cut. If the entire assembly is made of plastic the tubing can be bonded together with adhesive.

Included in the mechanism are hydraulic brakes with all the required implements. These drawings of hydraulic brakes are not functional as working models in the drawings but are simply representative of hydraulic brakes. Also, there is a twist grip throttle mechanism which throttle is cable and spring operated. The throttle operates by closing off the fluid flow diverted around the final drive impeller so the fluid flow is directed to flow through the impeller gallery: otherwise, with the throttle open the fluid flow is diverted around the final drive impeller and the wheel does not go around. The twist grip mechanism drawing is not functional as a working model in the drawing. These turnkey mechanisms may be purchased separately and are not intended as patentable with the patent application.

The ultimate final drive consists of a connecting chain loop that drives the drive train from the primary impeller input shaft. This assembly has a freewheel at the large chainring which freewheels clockwise (from the left side view) and locks counterclockwise, thus allowing the operator to crank at start-up while the power transmission unit assembly begins functioning. Once the applied apparatuses mathematics take over the ultimate final drive chain loop should apply force to the freewheeling large chainring exceeding the operator's force thereby reducing the operator's applied force to the crank while maintaining and increasing the vehicle's output performance with the throttle while still allowing the operator to crank. This is an unorthodox performance output means but it is mathematically probable and is the basis upon which human powered flight will be developed. This theory is understandable in the mathematics for Fluid Mechanics of Inertia in the power transmission application.

The only three prospects that will prevent this ideal from working up to specifications are suction resistance to fluid flow in the power transmission pump gears, a 90 degree direction change of the fluid flow in the gear pumps, and the fluid flow path in the impeller gallery. The fluid must change direction 90 degrees in the gear pumps twice, although the second time requires no pressure to be applied except to provide its escape.

The secondary power transmission must turn 16 revolutions per second at maximum speed, and the primary power transmission turns four revolutions per second at maximum speed. The depth of the secondary power transmission also increases the required fluid flow necessary with respect to the 90 degree change of direction. The primary power transmission is half as deep as the secondary power transmission but there are four times as many of them. Both the primary and secondary power transmissions should apply the same resistance force of suction continuously throughout the instrument's speed range due to the fact that one is twice as deep and turns four times faster and the other is half as deep and turn ¼ as fast but there are four times as many of them.

The fluid flow in the impeller gallery must follow a curved path from the inlet flow area to the draw suction around the impeller. This is constrained by the fluid flow passageways created by the design of channels to either side and above the impeller blade vanes. The remaining clearance of the impeller is limited to only a few thousandths of an inch. The likelihood that the fluid flow will follow the required path around the impellor is probable since it is the path of least resistance.

This instrument is also capable of coasting by which the throttle is released simultaneously (or not) and power is resumed by simultaneously applying cranking and the throttle at speed. The throttle will govern the cadence by regulating the fluid flow volume to pass around the output impeller thus allowing the secondary power transmission to speed up or slow down thereby increasing or decreasing the cadence while maintaining speed.

The operator must consume nutrients in order to apply power.

First attempt at human powered helicopter; contains secondary tubing

Second attempt at human powered helicopter unfinished andtoo heavy

Unfinished, Latest: needs control linkage, chain drive, tubing, throttle...

Chapter 7

Human Powered Helicopter Components and Attributes

The control of the lift blade vanes pitch is now being engineered. This consists of articulated lift blade vanes having ball joints with connecting rods which connect the articulation joining to ball joints mounted in a swash plate which swash plate is articulated around a ball joint in the center mounted to the main shaft to the lift blade. The swash plate is manipulated by control rods with respect to the forward and aft (pitch) blade vanes articulation control by synchronized bell cranks, and the left and right (roll) blade vanes articulation by independent bell cranks. The rotating swash plate is kept from twisting caused by the drag force of the blade vanes during lift, by fingers in the rotating swash plate that extend in to the central ball joint on four sides and by pins in the ball joint that protrude in to the main shaft ring made for them. There is only about 100 pounds of drag force being applied to the rotating swash plate through the connecting rods joining the lift blade vanes to the swash plate. The entire mechanism is operated by control rods and bell cranks to the operator's control stick which is an independent bell crank control lever on two axes with a universal joint at the two levers axes intersection which center will swivel about on a universal joint where the two axes of the levers' rotation intersect. The steering yaw control rotates through the universal joint to the yaw control mechanism to change the tail rotor blade vanes articulation (yaw). The operator moves the handlebars in a circular manner to control the lift blade vanes roll and pitch blade vanes articulation control and the arc levers at the base of the handlebars stem (at right angles to one another) operate the bell cranks that control the swash plate. The operator rotates the handlebars clockwise and counterclockwise to control the tail rotor blade vanes articulation.

The tail rotor blade vanes articulation is controlled in the same way as the previous design of human powered helicopter. The mechanism has not been changed much, only the design appears to be similar. The only drawback remaining in the entire assembly is the flexing of the connecting rods during operation since there now is only a single connecting rod where there were once pairs.

The Power transmission units and the impeller galleries are to be made of polyurethane surrounded by epoxy-carbon fiber with metal inserts for bearing races and containing the gears in the power transmissions. Any other component which may need a metal piece may be applied. Bolts will be inserted with threaded inserts in to the polyurethane and epoxy-carbon fiber for mounting and assembly. The airframe will include bearings to support the impeller gallery output shafts to prevent deformation of the epoxy-carbon fiber and polyurethane constructions, as well the secondary power transmission units will have bearings supported by the airframe. The primary power transmission units' chainrings will have bearings exterior to them to prevent deformation during loading. These bearings are placed at the chainrings where the applied forces are critical. The calculated rate of the impellers for both the primary impeller and the tail rotor primary impeller are four revolutions per crank circuit of revolution. In the previous design of human powered helicopter the rate of impellers revolutions was sixteen revolutions per crank circuit of revolution. Also in the drawings and in the calculations the lift impeller blade vanes profile area is calculated at ¼" square and not ½" square to calculate the secondary power transmission unit meshing gears teeth depth.

The center gears are to be made of titanium, not to say that the center gears need to be tough but that the output shafts need to be strong and are directly tied in to the gears. The center gears and output shaft are made as one piece. The planet gears may be made of aluminum plated with stainless steel. The output shafts for the lift blade and tail rotor are also titanium. The blade vanes hubs for both are titanium. The lift blade spars are titanium. The lift impeller is titanium. The tail rotor impeller can probably be aluminum but if it is titanium it will weigh less although it will cost more. The airframe is 6061-T6 aircraft aluminum but may have to be made of reinforced fiberglass to reduce the weight of the helicopter while hopefully not sacrificing too much strength.

To date, the chain of the chain drive is motorcycle chain, ½ inch width and ¾ inch pitch, which may be stainless steel.   The 72 tooth sprocket is a freewheeling sprocket freewheeling clockwise when viewed from the left side. This will allow the operator to crank the pedals crank while the secondary impeller begins to turn. By design calculations the fluid flow to the secondary impeller is greater than is required to turn four revolutions per crank due to the clearance flow volume and the secondary impeller should transfer that extra revolutions force to the 72 tooth sprocket to assist the operator in cranking. However, this may turn in to a runaway mechanism and a throttle will be need to be returned to the assembly. A relief bypass flow may need to be employed to control the runaway problem, if there is one. Sprocket spacing at the primary power transmission units is 10 ½ inches in two dimensions. Sprocket spacing between the 72 tooth sprocket and the primary impeller is approximately 165 inches to allow for chain slack so the chain doesn't break. Sprocket spacing between the 52 tooth chainring at the operator's crank and the flywheel also allows for slack so the chain does not break: however, chain slack in the flywheel drive should be as little as possible to sustain the maximum performance of the flywheel at the operator's crank.

In the drawings there are 36 tooth sprockets at the all the power transmission units. This calculation has been in error and their ratio to the center gear in the power transmission units is 4.5 : 1. With 24 tooth sprockets at the power transmission units the ratio at the center gears will be 3 : 1 which is what the original calculation were supposed to be at. Also the 25 tooth sprocket at the crank will be changed to a 24 tooth sprocket, leaving the 36 tooth sprocket at the crank. All these sprocket changes will affect the spacing between the power transmission units so that the chain will align on the sprockets. I haven't calculated the spacing between the sprockets yet and will not redraw this first drawing of the human powered helicopter but I am going to draw a new version of the human powered helicopter.

Getting the weight down of the helicopter is a primary consideration.

The human powered helicopter is being redesigned. The power transmissions are being located directly under the lift blade; the collectors from the primary power transmissions are being located directly forward of the power transmissions forward of the lift blade axis or rotation. All other components are normal as previously drawn. The aircraft has been shortened by 13 feet seven inches by locating the power transmissions vertically under the lift blade axis. Also, the lift blade primary power transmission units are on the bottom of the stack, the tail rotor primary power transmission units are lighter and are therefore on top of the stack of power transmission units located vertically under the lift blade axis. The primary impellers are both the same from the previous model's design. Their calculations have been adapted. Only the primary power transmission of the tail rotor has been changed with the secondary tail rotor power transmission because the tail rotor is moved from 40 feet out to 27 feet out so the calculations had to be redone and the power transmissions had to be recalculated. The primary impeller for the tail rotor has remained the same from the previous model's design. Hopefully still, the human powered helicopter will not turn out to be an elaborate ceiling fan. The success of this latest design will be on the lightness of parts. The appearance of the assembly design is representative of the parts' locations. The aircraft must be constructed to meet FAA certification for airworthiness and certificate of airmanship. The end result may not even appear to be the same constructed materials as the drawing portrays them. The airframe may have to be made of plastic because if it is aluminum it will most probably turn out to be too heavy. The output shafts however will have to be Titanium because of their lightness and strength in the places where they are applied, as will the impellers, because of the forces. Torque is very high and forces are great in the places where the Titanium is applied. A minimum of Titanium is ascertained in the design to keep down costs and applied where necessary while aluminum is applied otherwise. The assembly drawings of the human powered helicopter is just a rendering, the actual end result of the parts designs will be different once the values of applied forces in static equilibrium are evaluated for limits of tolerances on parts geometries for shear areas at absolute minimum elastic limit at applied maximum force and the aircraft is constructed to its minimum possible weight. This capacity however may take 100 years to file out of the original prototype (hone to exacting specifications), so I am not expecting miracles out of the original prototype, but it can happen if the right people and today's technology is applied.

This latest version of human powered helicopter has an error in it. The tail rotor impeller is too small resulting in applying too much force to the operator at the crank. As you can see I used previously designed parts to apply to the latest version of the human powered helicopter and the tail has been shortened by several feet increasing the force at the tail rotor without having changed the impeller diameter. Correct this error. The applied force at the operator is about 250 pounds at take off with it as it is.

Hopefully the human powered helicopter once it is finally designed will be light enough to include an electrical system including an alternator, a 12 volt battery, all the accoutrements necessary for driving the 12 volt electrical system and charge the battery, a voltage meter/charging circuit, and amp-meter, an altimeter, two transceiver multi-channel radios with output power/range control, running lights, landing light, strobes, oscillating red light on the tail, an interior light, and an air speed indicator, and circuit breakers. Not much else is really necessary for a human powered helicopter except maybe hydraulic fluid levels in the reservoirs and vacuum sensors in the hydraulic lines, with indicator meters. Warning lights for all the essential points would be considered convenient, and alarms would be convenient as well, as this helicopter is manually controlled there will probably be no need for a computer to monitor anything. Warning indicators should be sufficient.

I haven't finished this last attempt at the human powered helicopter either. It needs control linkage to the lift blades from the operator, the chain drive linkage, the remainder of the tubing which can be drawn using the other human powered helicopters I have drawn, reservoirs, and the self-propeller motor tubing sizes, and it needs handlebars, some other things, and a throttle. This throttle however may be linked between the suction of the primary impeller tubing and its head outlet tubing instead of between the inlet and outlet tubing of the lift blade impeller. Thanks. I will finish drawing this latest attempt to draw the lightest human powered helicopter of the three eventually.

Chapter 8
The Human Powered Helicopter

The evaluation of the power train assembly of the human powered helicopter is the same as for the Extreme Bicycle, only there are a few more evaluations for calculations that must be made for the human powered helicopter. The human powered helicopter has a lift blade and a tail rotor blade which are both powered by impellers. Their calculations are as follows: If the lift blade vane is an airfoil shape and has a straight depth resulting in a rectangular side view profile describing the location of the applied moment of lift is in the diagrams on the following pages.

Chord area equation, ½ of a blade vane profile area

Minimal helicopter lift blade calculations... (rough sketch)

Bell crank calculations, airfoil vectors in performance

Once the applied force is found for the lift blade impeller and the tail rotor impeller the same calculations apply to calculate the power transmission gears depths and the primary impellers radii to allow for the sustained operator's input force at the crank. If the force at the crank is out of tolerances simply apply the desired force in ratio with the calculated force in a fraction greater than one (1) and multiply that fraction to the radius of the impeller to equal the new radius of the impeller which will align the limit of tolerances with respect to the operator's force at the crank. The two forces between the lift blade and the tail rotor blade at two crank circuit-of-revolutions per second sustain the maximum operator's force at some desired altitude, say 10,000 feet air density, with respect to the weight of the aircraft (which is required to be minimum, trimmed at every possible place).

The human powered helicopter has Titanium components. The output shafts are Titanium. The secondary power transmissions center gears input shafts are Titanium. The lift blade impeller is Titanium. Components with very high torque force and small radius are Titanium.

The power transmission cases are made of poly-urethane encased in epoxy-carbon fiber with aluminum sleeves for the epicyclical gears and fasteners inserted. The impeller cases are the same materials as the power transmission cases.

The airframe is 6061-T6 aircraft aluminum. The power transmission mounts are welded to the airframe. All the particular components of the airframe are welded to the airframe in their respective places.

Final drafts have yet to be drawn. The specific tolerances of the individual parts have to be finalized before the final drafts can be made. The parts balancing has to align so the sum of the moments equals zero on the lift blade axis including the operator(s), which operators are all different and each human powered helicopter has to be designed for the variety of operators that there are.

Human powered helicopters will not be inexpensive until they are becoming mass produced and even then their retail price will be significant.

This human powered helicopter is designed to lift a 275+ pound person. The calculations for the lift blade apply 2,431.6 pounds of lift and drag to the lift blade at maximum output. Lift blade revolutions output are shown in the calculations. A calculation of torque applies force to the impeller inside the lift blade impeller gallery as shown in the calculations. The formula for Fluid Mechanics of Inertia applies the required volume of hydraulic fluid to account for the static load of force at the impeller times revolutions plus fluid flow through the relief passageways at static flow volume output at maximum applied force. This sum quantity equals the initial meshing gear teeth volume of the secondary power transmission. The same force at the lift impeller applies at the secondary power transmission. Note that all of the hydraulic forces are in SUCTION. Next, the clearance volume in the secondary power transmission is calculated at the same force, the exponent of the formula for Fluid Mechanics of Inertia is multiplied to the clearance volume and the product is added to the initial meshing gear teeth volume of the secondary power transmission. This is the final volume of the secondary power transmission. Note that all these calculations are made with respect to two crank revolutions at the operator pedals and chainring per second which is equal to maximum output performance, 128 cadence per minute.

The applied force is still the same at the secondary power transmission meshing gears teeth. Now, roughly estimate a circumference for the secondary power transmission impeller and calculate its torque using the force at the meshing gear teeth of the secondary power transmission center gear diameter (preferably the working depth diameter). The force at the impeller radius working depth of the impeller blade vanes is now calculated back through the drive train to the operator. Once the desired Operator force is applied divide a percentage of the desired force at the operator (say, 67%) into the resulting force of the calculations at the operator and multiply the quotient to the diameter of the impeller and the product should result in the desired diameter of the impeller resulting in the 67% required operator's force through the drive train at the pedals center axis. The 67% applied force will equal 100 minus the tail rotor force divided by the lift blade force times 100 equals the desired percentage of primary power transmission operator's force.

The applied force at the secondary power transmission is now calculated in torque to be the desired force at the resulting diameter of the secondary power transmission impeller diameter. Use the formula for Fluid Mechanics of Inertia to calculate the fluid volumes of static load volume of the impeller blade vanes times revolutions plus the relief passageways volume at static load exponent, and this will be equal to the primary power transmission initial meshing gear teeth volume. The applied force at the secondary power transmission impeller conveys to the primary power transmission and is the same force. Now calculate the clearance volume in the primary power transmission and using the force at the secondary power transmission impeller blade vanes and the formula for Fluid Mechanics of Inertia calculate the static volume of the clearance volume of the primary power transmission at the applied force using the exponent of the formula for Fluid Mechanics of Inertia and add the product to the initial meshing gear teeth volume of the primary power transmission initial meshing gear teeth volume. This will be the final meshing gear teeth volume of the primary power transmission.

The Tail Rotor

The applied suction at the lift impeller will make the helicopter want to rotate the airframe clockwise looking down from the top. Again, torque applies at the radius of the lift impeller with respect to the axis of the tail rotor. Keep in mind that all forces are applied with zero motion although motion is accounted for as revolutions in calculations. The same formulas apply with respect to the tail rotor blade vanes and with the lift rotor blade vanes although the tail rotor blade vanes will pitch to make changes for yaw control, increasing and decreasing or reversing the amount of lift plus drag force created by the tail rotor blade vanes. The entire force at the tail rotor (lift + drag) is used in calculations to determine the dimensions of the impeller and the power transmission meshing gear teeth volume of the tail rotor power transmission assembly exclusively.

The impeller for the tail rotor is calculated to turn the required output revolutions by minimizing the frontal area profile of the tail rotor impeller blade vanes while managing the secondary power transmission of the tail rotor impeller so the impeller of the tail rotor secondary power transmission and primary tail rotor power transmission are all kept limited to minimum tolerances. Varying the diagonal frontal area profile of the impeller blade vanes and the impeller's diameter will align the tail rotor secondary power transmission to minimum tolerances. The applied force at the tail rotor impeller (lift plus drag (lateral and transverse)) is calculated back through the drive train to the operator and added to the initial operator's force (for example: 33% tail rotor force + 67% lift blade force = operator's force). The sum of the two forces at the operator should be the desired operator's force. The diameters of the secondary power transmission impeller and tail rotor secondary power transmission impeller must be aligned to achieve the desired operator's force.

The sum of the two forces is applied at the operator pedals and must be within the operator's tolerances. Hydraulic fluid should be sewing machine oil or some likeness of thin lubricating oil of low viscosity, perhaps linseed oil but must not break down over time and not be hygroscopic. Most parts are to be made of elastic plastic or some other durable light material that is inflexible. The control linkage is a simple gear driven bell crank and connecting rod assembly.

The tail rotor spins counterclockwise looking at the tail rotor from the left side of the helicopter. The tail rotor mechanism consists of a ball jointed bell crank which turns a screw that operates a lever with a forked end that moves a slider (the internal component of which spins on the tail rotor output axle) that operates the blade vanes to pitch, whereby the lift and drag of the tail rotor are controlled and so yaw occurs. The tail rotor blades turn on an axle that is engaged with the tail rotor, and output impeller. The screw threads of the screw that operates the lever are left hand threads for the required blade pitch with the steering and control linkage.

The Secondary Hydraulic System

All of the hydraulics have reservoirs. The secondary hydraulic system operates by syphoning hydraulic fluid from the reservoirs. The purpose of the secondary hydraulic system is to prevent air from leaking in to the entire assembly at the axes of rotation through the grease seals. Since there is suction throughout the entire assembly during operation under load there is continuous suction at the axes of rotation at the grease seals and hydraulic fluid will be sucked in from the reservoirs through the secondary hydraulic tubing instead of there being air sucked in at the grease seals. By applying secondary hydraulic tubing to suck hydraulic fluid under suction force the possibility of sucking air in at the grease seals is eliminated. Hydraulic fluid is continuously replenished in the reservoirs by the continuous recirculation of hydraulic fluid in the assembly.

There are six reservoirs on this helicopter.

Pitch and roll are controlled by the operator leaning forward and backward and from side to side. There are four locations for counterweights at the rear of the helicopter. The counterweights are bullet shaped with a threaded end and hang down by their threaded end. The counterweights will probably be to offset the weight of the hydraulic fluid. A counterweight to offset the weight of the pilot has not been added at this time. No account has been taken for lift blade vanes pitch control at this time.

There is a throttle. The throttle is hand operated by a twist grip mechanism at the right hand on the handlebars. The throttle controls the flow of hydraulic fluid to the lift impeller by closing off the bypassing fluid flow bypassing the lift impeller gallery by channeling it through the throttle and allowing more hydraulic fluid to flow to the lift blade impeller gallery thereby applying more hydraulic fluid flow to the lift blade impeller gallery increasing the force of flow to the lift blade. The operator may crank the pedals and vary the cadence while operating the throttle and control the lift blade output revolutions and the cadence by manually operating the throttle, to be able to ascend and descend in a more controlled manner and at a more comfortable cadence.

The control of the lift blade vanes pitch is now being engineered. This consists of articulated lift blade vanes having ball joints with connecting rods which control the pitch joining to ball joints mounted in a swash plate which swash plate is articulated around a ball joint mounted to the main shaft to the lift blade. The swash plate is manipulated by control rods with respect to the forward and aft (pitch) blade vanes pitch control

by a bell crank, and the left and right (roll) blade vanes pitch by independent bell cranks. The rotating swash plate is kept from twisting caused by the drag force of the blade vanes during lift, by fingers in the rotating swash plate that extend in to the central ball joint on four sides and by pins in the ball joint that protrude in to the main shaft ring made for them. There is only about 100 pounds of drag force being applied to the rotating swash plate through the connecting rods joining the lift blade vanes to the swash plate. The entire mechanism is operated by control rods and bell cranks to the operator's control stick which is an independent bell crank control lever in two axes with a head tube at the two levers axis which axis will swivel about on a universal joint where the two axes of the levers' rotation intersect. The steering yaw control rotates through the universal joint to the yaw control mechanism to change the tail rotor blade vanes pitch (yaw). The operator moves the handlebars in a circular manner to control the lift blade vanes roll and pitch pitch control and the arc levers at the base of the handlebars stem (at right angles to one another) operate the bell cranks that control the swash plate.
This concludes the explanation of the human powered helicopter.

Chapter 9
The Propensity Of The Self Propelled Motor

As the rocket motor seemed to be self-propelled so the self-propelled motor would seem to propel itself as well. Thus, the mechanics for self-propulsion of motion in a motor is presented in a mechanism in these drawings; these drawings are placed in order of evolution:

Evolution of the self-propelled motor, 1st drawing (CAD software)

Second drawing (CAD software)

Final draft (CAD software)

The mathematics for the self-propelled motor apply with respect to the effect that multiple revolutions are applied in product (multiplication) with respect to the coefficient of the impeller gallery volume and the number of revolutions desired exceeding one (1) revolution of the impeller. By calculation the meshing gears area is divided into the impeller gallery volume less the impeller x the number of revolutions = the meshing gears depth. The meshing gears radius is less than the impeller blades vanes radius by a significant ratio. Preferably the meshing gears depth makes the meshing gears a square by side view profile with respect to the radius of the meshing gears compared to the impeller blade vanes radius and applied calculations for the depth of the meshing gears. But the square doesn't really make all that much difference, the object is to get the hydraulic fluid to suction freely to the power transmission throughout the assembly, from the head flow of the meshing gears to the reservoir, from the reservoir around the impeller blade vanes in the impeller gallery and back to the power transmission suction inflow meshing gears and around the gears to the head flow meshing gears in a continuous cycle, the excess head hydraulic fluid being bypassed by the throttle valve before it gets to the reservoir back to the suction lines allowing for acceleration and deceleration of the motor in performance.

Since the moment is in the flow force from the suction at the meshing gears inflow and the flow of the hydraulic fluid around the impeller blade vanes is applied in multiple revolutions with respect to the revolutions of the impeller's actual turning rate, the flow force on the impeller blade vanes applies torque to the meshing gears which applies suction force to the hydraulic fluid in the meshing gears which force is greater than the force at the impeller blade vanes because it is applied in torque and the meshing gears radius is smaller than the impeller blade vanes radius, the throttle valve will bypass the excess flow of meshing gears suction of hydraulic fluid from the head flow back to the suction flow standing idle sustained while closing the throttle will reduce the head flow to the suction meshing gear teeth and increase the suction through the impeller gallery and the head flow to the reservoir so more flow goes to the impeller in the impeller gallery applying more force to the impeller blade vanes increasing the torque and therefore the force at the meshing gears teeth and as a result creating acceleration in the internal working parts as well the reverse is true creating deceleration in the working parts by opening the throttle valve allowing the head flow to pass to the suction meshing gears teeth suction lines and not to the reservoir. The force applied at the suction meshing gears teeth is the same force that is applied at the impeller gallery port area through the hydraulic line, so there should be torque.

With all due respect this may be the only self-propelled engine that may ever work, if it works. The reason the author has concocted this concept is that it contains all the hydraulic fluid and does not spill any of it out like the designers of self-propelled engines of old spilled all their hydraulic fluid out in their drawings of self-propelled engines.

As you can see in the evolution of these designs the diameter of the tubing aligns, from straight tubing of one single diameter to tubing of the sum of the areas of the tubing in line with the exception of the area of the ports at the impeller gallery. The alignment of the areas allows the hydraulic fluid to flow without adding any additional force of acceleration from having to channel a sum of areas in to an area equal to 1 area/sum of the areas x (sum of the areas) = 1 area, which would cause the hydraulic fluid from the sum of the areas to have to accelerate in to the 1 area creating extra force to have to be applied at the suction meshing gears teeth reducing the efficiency of the performance of the self-propelled motor.

Now the tubing size is equal to the sum of the areas. The hydraulic fluid flow is constant. There is no acceleration force to add. The head force no longer has an added force from having to accelerate the hydraulic fluid. Deceleration of the hydraulic fluid would not have applied any useful effective force in the self-propelled motor because the force of deceleration would only have applied a load to inflate the draw force meshing gears with hydraulic fluid as the gears turned. Still, eliminating the acceleration of the hydraulic fluid eliminates the restriction of the forces of acceleration from a single diameter of tubing.

Suction is applied throughout the hydraulic fluid on the suction lines side of the system to the inlet port area at the reservoir just before the impeller gallery no matter what the diameter of the tubing is, the suction is still the same at the inlet port area no matter what the diameter is between the inlet port area and the meshing gears suctioning the hydraulic fluid. If there is a gas in the lines there will be spring/bounce in the suction force. The gas must be removed for the force to be solid.

The head side of the meshing gears only pushes the hydraulic fluid to the reservoir it does not have any pressure on it. Only the viscosity of the hydraulic fluid will apply reaction force in the meshing gears teeth as the hydraulic fluid is squeezed continually between the teeth of the meshing gears to be pushed to the reservoir. This squeezing effect is continual in the performance of the self-propelled motor of this kind, as is the suction flow of the hydraulic fluid continual simultaneously.

The dimensions of the impeller blade vanes can be changed. Its radius and depth can be varied to accommodate a limit with respect to a certain size of inlet and outlet port area. This size change however will affect the meshing gears depth. All the respective radii are taken into consideration. So, if you are going to make the impeller blade vanes big and fat you will probably have to change the radius of the impeller so the meshing gears depth will make the side view profile of the center gear a square when its depth is calculated. If the impeller blade vanes are big and fat the radius of the impeller will be really small if the side view profile of the meshing gears is still a square. Surely there is a point of diminishing returns on the radii which can be applied when designing this self-propelled motor. I will leave getting the most out of this motor for the future of mankind and their ability to develop machines to their state-of-the-art capability.

To calculate the size of the power transmission meshing gear teeth volume the size of the impeller has to be described by choice. The application of fluid flow volume to the impeller can either be by a multiplier, by a multiple of two, or by a force in pounds, preferably in a multiple of two simplifying the analytical procedure. In this case 24 is the multiplier used to calculate the gears depth. Without using the formula for Fluid Mechanics of Inertia and the density of the hydraulic fluid, calculate the volume of the impeller gallery less the impeller volume, impeller bracket gasket and impeller bracket and multiply by 24. Calculate the impeller blade vanes volume and multiply by 24 x 1.5. Divide the sum of the total volumes by the area of the meshing gear teeth of the power transmission, using one area and multiply that one area by the number of meshing gears. This is the final meshing gear teeth volume for a 24 multiplier power transmission. Then divide by two power transmissions. Fabricate gears (center and planet gears are equal depth) according to these calculations. Fabricate the power transmission casing to the same depth. A +0.007" gasket separates the power transmission casing from the power transmission plate. The power transmission plate contains the ball bearings for the center gear and the pilot bearings for the planet gears and the openings for the fluid flow tubing, and bolt holes for the power transmission grease seal. Assemble the power transmission. The grease seals fit over the center gear output axle and bolt to the power transmission plates with a gasket between them. Assemble the power transmission to the mount.

Assembling the impeller assembly

In this case apply a 3 inch radius to the impeller. This is the working depth of the impeller blade vanes. These impeller blade vanes are ¼ inch by ¼ inch in the profile view. The total radius of the impeller is 3 1/2 inches at the impeller blade vanes. The impeller is ½ inch thick at the impeller blade vanes but the blade vanes are only ¼ inch thick leaving 1/8 inch to either side of the impeller blade vanes for clearance volume. Apply a 3 1/2 + 1/16 inches radius to the outside of the impeller and extrude to depth to one side approximately 1 inch deep, this leaves 1/16 inch clearance above the impeller blade vanes. To the other side apply the same radius to an extruded depth of ¾ inch. The depth of the deeper side of the impeller will have to be adjusted as the impeller casing is fabricated. Fabricate the casing of the impeller to include 0.007" radius clearance (+0.014" diameter) for the impeller to fit into the casing. The grease seal and gasket fit over the center gear input axle, then the impeller casing fits over the input axle of the center gear and the bearing goes in and bracket fits over the end of the center gear output axle and is attached with an E-clip. The impeller brackets attach to the center of the impeller on one or the other sides with a +0.007" rigid gasket between the bracket and the impeller to keep hydraulic fluid from leaking out through the bracket at the center gear input axle. The bearing rides on the bracket. Clearance is machined in to the impeller casing for the bracket, gasket and bolts, and a space is machined for a bearing in to the impeller casing. The remaining bore through is +0.014 " the diameter of the bracket diameter that will stick through the impeller casing. The depth of the casing for the impeller will allow the fluid flow ports to be drilled out to allow for heavy diameter and heavy wall thickness tubing, this is in the event that if there is a given load to the self-propelled motor and the suction force increases the tubing will not crack. The holes drilled in the casing must have enough material to one side (the open side for the impeller to enter) so the drilling operation will not deform the material at its thinnest area to the open side. Therefore, the impeller depth to one side must be aligned to the plane of the impeller casing depth while the impeller blade vanes and clearance area and fluid flow ports centerlines must align keeping the +0.007 inch clearance all around.

The impeller grease seal and gasket go on the other center gear input axle, then the impeller plate fits over the same center gear input axle and the bearing goes in, and the other impeller bracket goes on the end of the center gear input axle and attaches with an E-clip. A 0.007" rigid gasket goes between the bracket and the impeller and the brackets are then bolted to the impeller. A +0.007" gasket goes between the impeller casing and the impeller plate and the two halves of the impeller assembly are fitted together. The same bores are taken for the bracket to the impeller on the plate side including the bearing. The entire assembly is mounted on the mount as construction is proceeding.

Assembling the tubing and self-propulsion

The elbow tubing is placed first at its respective angles according to the drawings. The straight tubing and elbow tubing are joined by rubber sleeves with hose clamps and the tubing are pressed as close together as possible to prevent an aneurism from occurring in the rubber sleeve at the gap in the tubing where the tubing comes together. The tubing is butted together. The tubing fits in to the collectors with rubber sleeves. The tubing is glued or soldered in to the impeller plate fluid flow ports. The throttle also is fitted with rubber sleeves. The secondary hydraulic tubing is fitted from the grease seals to the reservoir and fixed in place. Hydraulic fluid is drawn in to the grease seals from the reservoir through the secondary tubing as suction is drawn by operation of the impeller and with load as operation applies suction. Hydraulic fluid is recirculated back in to the reservoir. The throttle tubing is assembled the same way. The secondary hydraulic system does reduce the probability of power slightly but it is the only way to seal in the hydraulic fluid. The reservoir is filled, a crank is inserted in to the outboard end of the center gear (on the right side) and the crank is turned clockwise as the throttle is closed, hydraulic fluid flow is channeled in to the impeller gallery by suction and away from the throttle (throttle closing), the amount of force of suction being applied at the meshing gear teeth is the same at the impeller suction fluid flow port; there is a 3:.95 inch radius ratio of torque where the radii of the two forces are equal (between the impeller blade vanes working depth and the meshing gear teeth circumference). As the throttle is closed the flow volume to the impeller increases increasing the torque while the forces remain equal. The force of torque being applied would seem that the sum of the moments greater than zero is achieved. This device can be varied by applying the gears depth with respect to the radius of the impeller (changing the multiplier), the impeller clearance volume around the impeller blade vanes, and the blade vanes volume. Some other variables may apply like gear teeth count for smoothness and minimizing friction, but the author is unable to constitute a reasonable faction for why this self-propelled motor will not work except that self-propelled motors are notorious for not working.

Update

Lately the tubing diameter has changed. The areas of the tubing have been equalized so the sum of the areas of the six inlet ports of the collector are equal to the area of the outlet port of the collector. The related tubing area have all been made the same diameter. The only change in area occurs at the port area of the impeller gallery at which point the fluid flow is required to speed up anyway. This tubing size change relaxes the hydraulic load on the fluid flow. Still, hydraulic fluid flows to the inlet and outlet ports of the impeller gallery only with less haste and there is virtually no pressure or suction load compared to what is inside the impeller gallery. Pascal's Law of areas in hydraulics makes an equal amount of volume displace from six in to one out with equal area for both. I suppose the displacement distance ratio is 1:1 if the areas are equal.

Here is a problem: Suppose head and draw forces are equal at the meshing gear teeth at 7 + 7 = 14 at the radius of

.95. The radius of the impeller is 3 and has a force of only the one draw force of 7. 3 x 7 / .95 > 14 x .95 / 3. Even if there is no head force to the impeller there is rotation in the torque lever arm. All the head force is expelled in the reservoir. Even more probable is there is very little head force at the meshing gears teeth head outflow. Only the viscosity of the hydraulic fluid will introduce head force. Figure out some way to prove that this motor applies with respect to the sum of the moments equal to zero and I will admit that it doesn't work. No friction is included in this equation but will be included in the actual working model.

For reference, all the previous "perpetual motion" motors I have seen in my life growing up spilled all their hydraulic fluid out in their attempt to perform; my self-propelled motor's hydraulic fluid is all contained and none of it spills out. Also, this is not physics this is statics.

It seems impossible but even when the moving parts are held in static equilibrium and are kept from moving and the fluid forces are applied at their respective areas the torque equation still applies the sum of the moments greater than zero with the throttle fully closed. It may be required that the throttle be nearly fully closed for the machine to operate. With the moving parts released from static equilibrium and allowed to move freely and the throttle is closing mostly all of the way, and the parts are applying their respective forces all simultaneously, what could be causing the operation to fail? Discover the solution to this dilemma and resolve it by conforming the device to accommodate a solution to its operating principles. There must be some cause for why it won't work, the cause just hasn't been discovered yet. The solution for why it won't work has not been found. Only the areas ratio of the impeller gallery inlet and outlet ports and the meshing gear teeth inflow and outflow ports is left to question.

Even on the atomic level an atom or molecule of hydraulic fluid applying a force in the inflow and outflow of the meshing gears teeth (twelve atoms or molecules respectively (twelve inflow and outflow ports individually at the meshing gears teeth)) will apply six molecules or atoms at the inlet and outlet ports of the impeller gallery at the same force that is being applied at the meshing gears teeth and the sum of the moments will be greater than zero. Remember that the equation being applied to the impeller blade vanes is the equation for fluid mechanics of inertia: hydraulic fluid density x blade vanes displacement volume x 2 raised to the power of the quantity of the applied force divided by the quantity of the hydraulic fluid density x the blade vanes volume close quantity, minus 2 to the nth power, then divide by 2 to the same nth power, then close the exponent quantity and subtract the exponent quantity by the same n value, equals the hydraulic fluid density x the blade vanes volume x 2 raised to the power of p, which product is equal to the applied force in pounds. The applied force is the same force that acts in the meshing gears teeth, and if it is double for inflow and outflow for it to be the same force at the impeller gallery inlet and outlet ports simultaneously the equation for the sum of the moments is still greater than zero. The calculation for the meshing gear teeth volume is taken from the calculations from the impeller gallery volume to the meshing gears teeth volume as written at the beginning. This is complicated and the subject is beginning to go around in circles, no pun intended. Good luck. It is proving very difficult to make this self-propelled motor not work.

And, once again, the author is financially incapacitated and cannot function financially as an inventor: therefore, the self-propelled motor is being submitted for copyright. Thank you for your interest in the author's ideas and discoveries. Perhaps one day these ideas and discoveries will serve their useful purposes. No financial progress can be made. The author is sickened by the prospect of not being able to make any financial progress.

Even if the ratio of the sum of the impeller gallery inlet ports areas divided by the sum of the large areas of the four collectors is multiplied to any force and then multiplied to the ratio of the two radii in reciprocal and then one of the two is subtracted from the other while the fulcrum is the axis of rotation of the center gear, still the sum of the moments do not equal zero. Forcing the sum of the moments to equal zero is proving difficult.

Pascal's Laws show that a large area will sweep a short distance a particular volume and the same swept volume will sweep a long distance through a small area. The radii of the two areas in this case are in consideration: the port areas of the meshing gears teeth and the impeller blade vanes. Is the applied force to the impeller blade vanes diminished because the impeller gallery inlet and outlet port areas are smaller than the sum of the meshing gear teeth areas? Where did the applied force go? Was it dissipated somewhere? If so then where is it? If not then the applied force applies at the impeller blade vanes volume, area doesn't matter. Apply the formula for fluid mechanics of inertia, force is applied at the meshing gear teeth to the hydraulic fluid in static elastic equilibrium in zero time present simultaneously with respect to the simultaneous applied force to the impeller blade vanes volume on both radii. Notice how many variables can be changed: 24 can be changed to 16 or 8 or 12, these variables have been mentioned already, and the whole mechanism can be redesigned to accommodate the change in variables. Close the throttle and crank the crank, the self-propelled motor should start, if it doesn't I don't know why not. Keep trying.

Chapter 10
Gear Design Strategy
        A necessary component in the construction of the human powered transportation means is the sprocket. The following drawings give a quick overview of the design parameters for drawing sprockets on CAD software. It's easy so don't be too concerned. Gears need to be designed for inside the power transmissions, and sprockets need to be designed for the chain drive:

Gears design parameters

Sprocket Tooth Design Parameters For CAD, 28 tooth sprocket
Some techniques exist in the designing of a sprocket tooth in CAD but it is fairly simple. It takes practice but is not difficult.

Chapter 11

The Exploits Of Human Powered Flight

The problem with having a gasoline engine, while flying a helicopter is running out of gas and not having a place to land. A human powered helicopter is not like that, the problem with the HPH is you will run out of strength and all you will have to do is set down and wait and your energy will charge back up all by itself. You may also need to eat something but mostly just resting for a little while is enough to catch your breath and get your energy back to further your journey.

The prospects of going after the indigenous psychoactive intoxicant wildlife (i.e. the marijuana plant, the psilocybin mushroom, the peyote cactus (mescaline), the coca tree (cocaine), and whatever else I can't think of) is inevitable. People on human powered helicopters will scour the countryside for these plants and the Earth would be stripped bare of them were there no law enforcement case providing officers with human powered helicopters to prevent such a catastrophe, when actually the law officers will probably use the human powered helicopter to go in search of these indigenous life forms to destroy them themselves. Once the human powered helicopter comes of age mankind will seek out and lay waste to the wilderness of its indigenous psychoactive wildlife. However, this is not a bad thing. Once it is ascertained that the indigenous intoxicant wildlife is in danger of becoming extinct the Government should affect the endangered species act or whatever is necessary to protect the indigenous psycho-active intoxicant wildlife, which is counterintuitive according to the present state of intentions of the present Government.

Otherwise they will just wait and let man extinguish the plants from off the face of the Earth. Mankind is not smart enough to think before he acts when he craves these substances, dead or alive, especially in the United States where there is much abundance.

Human powered flight will make it possible to raid farms for food, although it would seem silly to form a raiding party with human powered helicopters since a human powered helicopter will be quite expensive and it doesn't seem likely that anyone who can afford one will be suffering from any kind of starvation prospects and would have no need to form a raiding party: still, in the future raiding parties may occur from natural disaster or some act of God. Farmers will need huge field sized nets to cover their crops so that individuals on human powered helicopters coming to raid their farm cannot swoop down and gather from the farmer's fields. Farmers shall not be allowed to shoot people however, they will only be within their means to call the police. The mesh of the nets will have to be fine enough to keep human predators out of the harvest, cutting through the nets covering a farmer's harvest could probably be a more serious offense, and getting caught one could do time, pay a fine, or both, for cutting a farmer's harvest cover net. Telephone lines will also be a bother. What will become of the telephone lines and power lines and electrical wires running between the telephone poles in and between cities and towns? It would seem feasible to string a fiber optic line from telephone pole to telephone pole that light up in the evening and stays lit all night so that the locations of the telephone poles and wires are obvious to pilots of human powered helicopters. The event would have to be funded by indiegogo or kickstarter or something like that. The military bases will have to do similarly and string a fiber optic light line around the military base to keep out the non-military public, or think of some other kind of means to deter people from flying in to a military establishment unawares.

The first human powered helicopter will have to obtain permission from Congress to fly over, into, through, upon, and under Federal Land Open To The Public to be able to leave a private restricted area in the first place.

Homes will have their garages built on their roof to accommodate the availability of human powered flight in the future. City skyscraper structures may have landing platforms built outside the offices of individuals for the purpose of convenience of access to one's human powered helicopter and/or office.

Flying this human powered helicopter will be elementary. Considering the modern day traffic or automobiles on interstate highways and during the rush hour, they are all bunched together on a single strip of pavement. Human powered helicopters will free man from the pressures of city life and grounded traffic opening up the sky with all of its innumerable altitudes and directions which can be taken on a whim. Establishing freedom in the sky for mankind in this country will require the Government yielding to the prospect that in time scenic routes will precipitate out and can be adopted as major throughways. For there to be determination to control the trajectories of our flights by imposing on man to dress and cover in flight, or to fly a designated flight path laid down before a scenic route has precipitated out, or to cause single file flying, bottlenecking et cetera, will need to be outlawed or prohibited in some way thereby eliminating traffic jams and a major cause of accidents in the sky. The deregulation of human powered flight should be a Right (as is prohibition) established by Government to free the people from catastrophe and so that our wings can never be taken away from us by any justification of justice. Owning a human powered helicopter should be a Right under the Fourth Amendment, as a keeping and bearing of arms and a rule needs to be applied that prevents the police from the confiscation of a human powered helicopter as a weapon is confiscated during an arrest.

As the nobleman goes along on his way, so the nobleman shall be free on his journey according to the Scriptures of the Holy Bible. Traveling in to wildernesses on a human powered helicopter should not be an excuse for law enforcement to pursue after the individual with the intent of apprehending them on any grounds without any probable cause, preponderance of evidence, burden of proof, or eye witness of a crime having been committed by the operator of the human powered helicopter only by the suspicion that an operator of a human powered aircraft is a drug addict. Appearing to not be doing anything does not justify apprehending someone in a National Park and imposing on the person to have to serve time in a mental hospital because he looked like he wasn't doing anything, in the National Park, should be investigated and set right. Since when does appearing to be doing nothing justify imprisonment in a mental hospital? Occupying one's free time in a National Park by one's self is not a crime, nor is it blatantly insane, or any intent to commit a crime and does not justify so much as even an approach by a law enforcement officer or Park Ranger. Flying to the mesa of a chimney in a human powered helicopter out West would deter any such approach, and sit there and appear to be doing nothing for the rest of the afternoon, you probably won't get approached by a cop on the ground up there and being taken in to custody and carried off from the top of a mesa tower, to serve time in a mental hospital, may prove difficult.

The objective by Government securing the sky as a National Preserve renders the requirement of a radio transceiver with multiple transmitter power settings which can be used to communicate short distances as well as long distances by the flip of a switch without shouting across the sky to another operator or shouting at all, because it is silent up in the sky except for the wind blowing and rain and thunder and the only shouting to date is the birds shouting to one another, birds are always shouting anyway, and the preservation of the silence should be kept secure, the sky should be kept as wild for future generations as it is now. Painting the clouds with food coloring should also be restricted to special occasions and the clean water act should take a precedent in case of overenthusiastic cloud painters making the water taste bad eventually, and change color from clear to a mucky brown from the mixture of all the different colors of food coloring.
It will be impossible for the operator of one human powered helicopter to reach out and touch the operator of another human powered helicopter, as you can determine by the

allowances of the blades locations during flight when approaching another human powered helicopter.

Still, it is not a wise decision to sow seeds while aloft in a human powered helicopter unless farming on one's own land, with the exception of sowing seed balls. Leave sowing wild seeds to the birds. Birds love to eat seeds and expel seeds, and there would be plenty of seeds to be sown if mankind had not got it in his head to lay waste to the indigenous foliage life forms over all the Earth in the meantime, on his human powered helicopter. Sit there on your sack of seeds and just smile and wave. Then there is the issue of where to land: in National Parks manmade "trees" can be built in remote places that will support human powered helicopters as landing pads and instead of landing on the earth and causing any erosion they can land on the "trees" provided for them: otherwise, it probably would be best to land on rock or concrete to prevent soil erosion in the wilderness, despite the fact that the human powered helicopters will probably only be able to land on flat surfaces including the water if so equipped. These "trees" will be made of materials and construction designs to support human powered helicopters like landing on fixed and rigid leaves. With pontoons you will be able to go fishing, and fly home to your lovely family and fix what you caught to eat. Getting chased by the cops will change too. From once long ago there was running and trying to get away being pinned to the Earth by gravity. Now there will truly be flight, and fight or flight will change to fight and flight, and eluding pursuit will be something to deal with. Dog fights may erupt on rare occasion between the assailants and the police.

The Holy Bible makes a reference to a holy multitude coming so dense that the light of the sun was darkened and the sound of their wings was as the voice of God. Should it be possible for everyone in this country to own a human powered helicopter then a mighty fighting force will we be in the event of belligerence, should the military need assistance. As well, human powered flight will make mankind capable of migration in the event of a solar system-wide cataclysm.

There should also be outlawed shooting wildlife from a human powered helicopter, including coyotes. The author does not think anyone should shoot any animal from an aircraft, how unfair is that. The author believes hunting from an aircraft is already outlawed in the United States but I don't know for sure. The author has seen a video of shooting coyotes from ultra-lights or something like that years ago on TV. The fact is the author cannot even fly the human powered helicopter that he would build even three inches off the ground without breaking the law these days. His human powered helicopter design would be an unregistered and uncertified aircraft in the United States' air space once it was built and would be subject to confiscation and he would get six months in jail if he allowed myself to be caught flying his human powered helicopter in the United States off of private property. Which is where deregulation of human powered flight is adamant, and human powered helicopters should be seen and privileged as an environmentally non-invasive contrivance of aircraft capable of sustained flight, and not requiring a license to operate. In which case you will probably need a good life insurance policy because these are the beginnings or the 21st Century, not the

22nd Century, and human powered helicopters are just about to be born in to this world and one will not be the most technologically advanced new thing since the model-T 100 years ago was not a speedy Indy race car of the 21st Century, so give leeway to the effect that 100 years have to pass before there will be the most technologically advanced human powered helicopter the world has ever seen.

Up the chimney will be a game: fly to a high altitude and cease cranking, let the human powered helicopter plummet to the Earth, and with precise enough timing engage cranking just in time to slow to a stop and touch the ground with the landing gear. Children will get good at this game. It is like driving the tractor when cutting the grass, practice, practice, practice.

In the most probability the favorite altitude will probably be to fly just above the tree-tops. This will facilitate fewer injuries from falls should a human powered helicopter fail to perform for some reason. A fall in to the trees may cause less serious injuries, and falls from low altitude as well will cause fewer serious injuries.

Human powered helicopters may not be user serviceable except for greasing bearings, adding hydraulic fluid, making minor changes to the operator's compartment or changing a chain, and some more serious replacements of some parts. Most of the human powered helicopter is not user serviceable: however, the human powered helicopter is designed to be maintenance free and should give more than a lifetime of good service, with proper care.

Human powered helicopters also present the formidable capability of spying on groups or individuals. Going in seeking of "where it's at" is also a formidable prospect for the use of a human powered helicopter. Catching a boyfriend or girlfriend cheating will be a capability of human powered helicopters. The flocking together of individuals will occur. It will at once be possible to see soul mates from a distance in the air, and flocks of people may form in time. This will also facilitate that these individuals will accumulate in neighborhoods as companions, or extended "families". Extended families, akin to flocks of birds, would more probably be the wiser organization of flocks of people rather than organizations of large flocks of "friends". Flocks of "friends" would be more likely to be disorganized and leave a trail of infractions in their wake.

Once the neighborhood has been established, then it will be possible to design cities and towns or communities around the human powered helicopter, with spokes of taxiways radiating out from a central runway. Houses, homes, cities etc. may be built on these taxiways: however, if HTOL human powered aircraft are not employed then with landing platforms on the roofs of houses there may be no need for taxiways or a runway for human powered helicopters. It may even be possible to have a job in another city should the human powered flight prospect prove fast enough to get to work with if one lives in another town from their workplace. However, in the meantime parking will be at a premium for human powered helicopters in parking lots.

Celebrities will have to carry umbrellas if they don't want to be recognized from the air. Nudist colonies will have to cover up. One day there will be a law that you cannot fly a human powered helicopter unless one flies wearing no clothes, but then this may be in extreme circumstances of the necessity to have to re-populate the planet should the human population drop to, say, twelve individuals or some critical limit where people are dying faster than they are being born and the population will become extinct in a matter of, say, half a generation. No one can tell the true circumstances of the Revelation To John unless they are Saved by The Grace Of God and can truly interpret the scriptures with reality. Who knows? Maybe there will be a time in the future when the population of the Earth is in danger of becoming extinct in a short time. Until then, this wearing-no-clothes business is just foolishness for me. Perhaps it is true, but how can I tell the future?

Nevertheless, putting restrictions on human powered flight, certainly for human powered helicopters, will only embitter the population, like trying to make them stop drinking coffee, with some sort of jurisdictional restriction.

The human powered helicopters should come equipped with a GPS, an altimeter, a turn and bank indicator, and a compass, and a two-way radio all incorporated in to a 12 volt DC electrical system with an alternator for power and all the necessary implements of 12 volt DC electrical systems with all loads running in parallel. This should be sufficient equipment to operate a human powered helicopter in Federal airspace open to the public. Once again these human powered helicopters are designed to be pilot-license-free and deregulated so mankind can enjoy the freedom of the skies without the burden of regulations and be free as birds to fly in the common air, unhindered by any persistent influence of justice during their traverse.

Speaking of which, there is the invasion of privacy issue, should one appear in the sky where another is in anticipation of their privacy being kept confident: therefore, privacy (in Kentucky) is lawfully maintained at an arm's length-to be to the tips of one's fingers as far as one can reach, and should there be any dispute as to whether one is "invading one's privacy" by appearing somewhere over the horizon, it shall be noted that in Kentucky privacy is limited to an arm's length and should any dispute arise from a "spy" anticipating the capturing of views, photographs, or whatever kind of media a record can be made on, of some individual, group et cetera, so as to ascertain facts-on-the-ground that the limits of privacy do not extend "as far as one can see" but to the tips of ones fingers as far as one can reach. Therefore privacy does not extend in to space as far as one can see, nor does ones property line extend vertically to the outermost reaches, but it is only an arm's length to the tips of one's fingers as far as one can reach. One should then be able to come in to view, approach, land and walk up to the suspected subject, or suspect, and stay just outside the full reach of their fingertips and NOT BE INVADING THEIR PRIVACY and take pictures, make recordings, or just look at them, just to get the wisdom and understanding that one should choose otherwise another mate, gather information or whatever is required. As long as one does not come inside of the arm's length extenuation then privacy is not invaded. This arm's length law needs to be established throughout this country in the event of human powered helicopters and the capability of being able to ascertain wisdom and understanding in the light of one's relationship with another. Breaking off a relationship on the pretense that you succeeded at finding your mate cheating, because you have a human powered helicopter and it gave you the capability to spy on them, is every reason to establish an arm's length privacy law nationwide and to outlaw privacy being "as far as one can see" (in any direction), because the mate will use the defense "as far as I can tell, or as far as I can see, or as far as I can understand" or such like, to establish their privacy in the courts, and the judge will need to establish that the "as far as…"rule, anything when it comes to privacy is that the realm is limited to within the tips of one's fingers as far as one can reach: as well, preferential treatment (or mistreatment in the case of a male defense witness) for the female prosecution witness by judges, and the defense council, in courts needs to be outlawed with fines or imprisonment or both and should be investigated.

Vulcanization of property to human powered helicopters should be limited to any three points on the property, whether it be a building, growing structures, vehicles, fences, persons et cetera, and the lines that join the three points creating an assembly of areas covering the entire property inside the property line. And the consolidation of properties between neighbors to increase the size of the vulcanized perimeter should be restricted.

Considering the prospects of one stealing a human powered helicopter: once before from the previous design it was not possible because the device was constructed to tailored tolerances for each individual, but now the aircraft is sustainable in flight by articulation of the blade vanes and some variation in operators weights can be accommodated for. So it may be possible for someone to manage to perpetrate theft of a human powered helicopter and make off with one if their weight is not too significantly different from the owner's. The human powered helicopter will need some kind of lock.

This human powered helicopter would not be too purposeful for use in the military although it may be useful for moving troops in large numbers over great distances in a shorter period of time than presently accommodated for, it is vulnerable to attack since the powerplants are made of Styrofoam and epoxy-carbon fiber. The operator is exposed to raw gunfire and has virtually no protection besides body armor. This aircraft may prosper for military means some distance behind the front.

There is also the prospect from the Holy Bible that the high places shall be made low and the valleys shall be raised up even with the plains. This may be a reference to human powered flight.

As for farming, the human powered helicopter may be used to lift somewhat heavy objects and carry them distance, or do some heavy lifting of sorts. Also, as with minor heavy lifting the military or Department of Transportation may be able to use the human powered helicopter to go on search and rescue operations. The human powered helicopter may also be used as a stealth aircraft because it may be very quiet. The Corpman may have use of the human powered helicopter in combat, with the help of a GPS and the coordinates of injured soldiers.

Going camping will now be a strategy curse for the camper-outer. The police will be like Were-wolves on human powered helicopter campers because of their likelihood of carrying illegal drugs despite the fantastic corporate reality that human powered helicopters will cost probably $280,000 for a new one easily and that casual consumers of controlled substances won't be able to go human powered helicopter camping because they won't own one. But who will go human powered helicopter camping anyway? Human powered helicopters are designed with carrying capacity now and practical camping is within the means of human powered flight. So beware of the police cracking down on you for using drugs if you have a human powered helicopter. Be aware thou owner of thine human powered helicopter, thou art a drug addict, and the police are going to make you know this, so be prepared to be searched and taken into custody if you are brought down by a police officer, and to spend the night in jail and to have your human powered helicopter impounded or confiscated. Therefore, the Rebellion: The Rights of the People to Possess, Own, and Operate human powered helicopters in Federal airspace and between the ground and their surface ceiling without fear or interference from law enforcement in that an individual shall be free on their journey (this implicitly includes California, without exception and without exclusionary rules applying, California is to be disciplined in its mistreatment of interstate traffic and corrected). Also, The Smokey Mountains are to be managed by Federal Authorities in the careless mistreatment of interstate traffic and corrected where human powered flight is prejudged as a means of trafficking in controlled substances or simply as a means of transportation for a drug addict when an individual on a human powered helicopter is seeking to engage in wilderness adventure.

Any location of public interest, mainly national parks and forests, recreation areas et cetera, where individuals are found to be being mistreated having been perpetrated for offenders for being occupants of human powered helicopters will be investigated and the suspected department of law enforcement subject to disciplinary action and corrected if there were no charges for drug offenses and all the subjected offenses were a result of the officer(s) having apprehended the subject(s) and all the offenses were ex post facto, e.g. concealed deadly weapon, public intoxication, assaulting a police officer, resisting arrest, none of which are drug related, and all are after the fact, since the police office arriving on the scene is more than likely after an individual on a human powered helicopter for drugs than for a pocket knife, or for having had one beer, or for talking with their hands, or for struggling because the officer just broke their arm having had them in a hammer lock. Nevertheless, the fact remains that the motive for the mistreatment has to have been drugs related, and finding out whether or not there are any drugs during the scene during the arrest is paramount in the discovery of the motive during the investigation. No police officer shall be left unturned.

If you leave the human powered helicopter alone long enough spiders will make their webs in it, birds will make their nests in it, the paint will flake off, it'll rust, it'll rot, it will leak, it'll get wet in the rain, snow will collect on it and ice will hang down off of it. It'll collect dust. The battery will go dead, branches will fall on it. Mud will get splashed on it. Parts of it will fall off or get stolen. It will sink in to the earth and begin to lean over. Eventually it will fall over and no one will pick it up neither can it pick itself up, and after a long enough time it will turn to dust. It is not a God. It cannot see or hear or speak or work miracles. It attracts lightning it does not send it forth. It can be broken into pieces and smashed to smithereens. If you ask it a question it does not respond because it cannot answer. It is lifeless and does not even know what it is. It does not know anything. It is a mechanism, a machine, which man operates with his mind and body. It itself can do nothing. If something breaks on it while you are in flight you could die, it cannot save you it has no powers of self-preservation. Do not worship it. It is made by craftsmanship as are all machines. Human powered helicopters are not Gods, or God, or any God. They are machines and are to be cared for as machines and not neglected lest they fall in to disrepair for they age as mankind ages because of the imperfection of sin and their days are numbered. This human powered helicopter is however a manifestation of The Will of the Christian God (I am), Jesus the Son of God, and The Holy Spirit having used a mere human to work as His conduit for doing His Will. The Human Powered Helicopter however is not yet manifested in physical form, only in design concept reflecting that mankind is not ready for human powered flight at this time, but soon. The fact that the author is financially unable to accomplish the construction of a human powered helicopter leads him to believe that he has much research to perform still and many months of study as well because there are still lots of loose ends in the development of a practical human powered helicopter.

I hope you enjoyed this manuscript of how to get power from a bicycle. There are still many details which are not presented here: the precision alignment of parts, their close tolerances, how to contain the hydraulic fluid while it is under extreme suction force while it is not really but it really is (it's difficult to explain), shear force areas of impeller blade vanes and input and output shafts, there are so many calculations. The formula for fluid mechanics of inertia solves them all using the cubit and the fold and the base formula. Practice practice. Apply this formula to everything that moves and has force and there problems will solve. Good luck.

Richard Chastain

To Contact the Author Send Your E-Mail messages to:

Rick.l.chastain@gmail.com

In The End There Was

How To Get Power Out of A Bicycle

By Richard L. Chastain

A Manuscript

(Unedited By Any Publishing Company)

A Variety Of Individually Written Manuscripts On Such Subjects As A Human Powered Helicopter, A Super-fast Bicycle, The Self-Propelled Motor, And The Social Impact Of Human Powered Flight, Including How To Get Power Out Of A Bicycle.

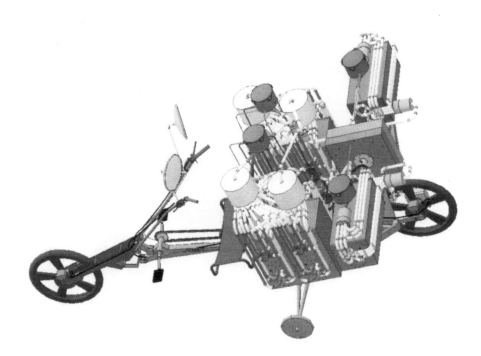

200 mile and hour bicycle

Introduction

   This book is not exclusively for highly educated people in the fields of engineering and science and does not exclude graduates of the elementary schools, although it does help to have an Associate's Degree in Mechanical Engineering Technology to start with. If one is able to add, subtract, multiply, and divide then you can do most of the arithmetic problems in this manuscript. You may learn to do some "algebra" and trigonometry, if you don't already know how, which will have to be agreed upon by the elementary school graduates as to the algebras' rearrangements of variables, also some statics, since elementary school graduates may not be too familiar with proportionality and ratios or equations for mobiles, bridges et cetera, certainly not equations for airplanes. For example: Crossing the equals sign with a variable makes a denominator a numerator and vice versa, and a plus sign a minus sign, a square a square root, and such like reciprocal-like stuff with respect to crossing the equal sign. Also there is the prospect of + + and - - (minus minus) which are both positive and I forget why I needed this though, it was necessary to eliminate a negative solution where a positive solution was required in some statics equations.

   The arithmetic is consistent, so don't worry too much about it being difficult. Only the variables (x, y, n, D, C, p, q, u et cetera) will either change or remain the same, while their identity will remain the same. This may seem confusing at first, but the variable for an identity can be different, the identity just has to be consistent with the variable in the formula. The same variable will most probably be used for a different identity to solve other problems with other equations, so the necessity to have to use the same variable exists because there are only so many letters in the alphabet and there are more than 26 identities. You don't have to memorize any of these variables, but the identities are necessary to work the problems whatever variable you choose to use in the equations. Subscripts are used as well to identify variables. Variables are also used for superscripts. Variables are also used to identify bases with superscript variables.

Identities have "units", which are like American Standard Units: pounds per cubic foot, and cubic inches, and cubic inches per cubic foot, and pounds, and things like that. Units, with respect to their "per" identity is the divisor (/), a forward slash, with a e.g. "pounds" in the numerator, a " / " and cubic foot in the denominator so that the relationship looks like this: pounds / cubic foot, "Pounds per cubic foot". The is the equation using "units" only in the formula for Fluid Mechanics of Inertia in the exponents' denominator quantity and the units are cancelled out: (pounds / ~~cubic foot~~ x ~~cubic inches~~ / ~~cubic inches~~ / ~~cubic foot~~) = pounds. Notice the "units" cancel because the denominator inverts and multiplies to the numerator: pounds / ~~cubic foot~~ x ~~cubic inches~~ x ~~cubic foot~~ / ~~cubic inches~~ = pounds.

Then there is the prospect of the dimensionless number, which is required to resolve the exponent of any base, in this case base 2. Pounds / pounds = a dimensionless number when Pounds, in the numerator, is > pounds in the denominator. > (greater than) is a standard symbol.

What started out as an opportunity to speak in front of the class on everything I know about bicycles has turned pretty much in to everything I ever wanted to know about everything I ever wanted to know anything about. Everything I know about bicycles started as a flopped opportunity to speak to the class, when after 30 years or more it has become engaging mathematics of human powered flight, interplanetary ballistics, space science, and rocketry, and using its basic formula to resolve every possible problem in space in force and motion in base 2 using the fold and the cubit (denominator quantity and coefficient to the base two and its exponent (the fold)) for which there must be a conclusion: ***the fold is the exponent of the base 2, and the cubit is the entire specific denominator quantity of a fraction*** (This Association Made With The Base Two Exponent And The Particular Denominator Quantity *Is A Fact* and should be acknowledged as such in tables). I will attempt to explain the performance of the basic equation of Fluid Mechanics of Inertia and apply it to useful functions of trade in today's world: e.g. a jet engine, a helicopter, an airplane, a bicycle, a rocket motor although a rocket motor has nothing to do with human powered transportation it still figures out when it comes to base 2 mathematics (I may not actually know how a rocket motor works but I have drawn a rocket motor concept which may be feasible), and such like devices as they may come to mind.

I write this manuscript for The People who have been left in the dark about everything that makes airplanes fly and helicopters fly and anything else a man or woman can get into and operate like a ship or a submarine, or a truck or a rocket ship and for the benefit of mankind, and for posterity, for those of us who have been kept ignorant of these things for so long, and to defy those who have kept these things secret from us all our lives, i.e. the aircraft industry, educational institutes of aircraft and spacecraft learning, the military, the space industry, and anyone else who would claim to know these things and for those who publish error and false knowledge telling us "the air travels faster over the top of the wing than it does under the bottom of the wing" (which is knowledge in error) when that is all they teach us, and write things after this manuscript is published and perpetrate themselves to be superior in their judgment rather than humble themselves having not written of these things until now for they knew what we were told by the ones who would have us know nothing and did publish information that they were taught or told to publish without knowing whether it is right or not assuming that it's right and in most cases making it far too difficult for the common person to understand.

So that everyone may benefit from my understanding of base 2 mathematics I will endeavor to exhaust the topic of "Fluid Mechanics Of Inertia – The Science Of Mechanics (the engineering of equations)" to the best of my ability in this writing and since it is basically simple mathematics I suspect that when this manuscript is completed that the books that will be left to be written will be so simple that a child can write them. I hope you enjoy this manuscript.

The only shortcoming in this manuscript is the engineering of compound-complex curved aerodynamic shapes to be used for aircraft. The means of assembling these shapes in to usable working models has not yet been discovered. How to construct a working model by assembling a variety of compound-complex curves in to a solid body shape and be able to calculate its geometric center because the shapes are calculable is not yet determined. Thus there remain all the formidable equations remaining to be calculated to find the center of gravity. Because these compound-complex curves shapes are calculable their subdivisions calculations should be simplified. Having to solve all these equations longhand with paper and pencil and arithmetic using a computer aided design software program to assist in the drawing of the simple to complicated compound complex curves shapes only facilitates the affect that numerous problems must be solved and proven successful before any interest is taken in the development of computer aided design software capable of solving the problems so we don't have to do any, or much, paperwork. Otherwise the problems get ignored and no progress is made in computer aided design (CAD) software programming, the likes for which this mathematical progress is probably incompatible with binary computer programming anyway and a whole new CAD software program will have to be designed for it.

Chapters are short and to the point. Some excerpts from other writings have been added for additional information.

Additional information can be found at www.fluidmechanicsofinertia.wordpress.com, keyed directly in to your browser.

Chapter 1

The Spacecraft

The moment the patch area of contact between the tire and the pavement makes contact with the pavement and the operator mounts the bicycle and begins to apply force to the pedals mathematics are set into play. There are mechanical action and reaction forces, and mechanics of fluid properties that act and react between the bicycle, operator, the equipment on the bicycle, and the surrounding atmosphere at its particular air density for its specific altitude, not to mention inclines and declines with gravity on slopes, all of which can be calculated statically or without motion while the bicycle remains in motion, and including friction which is minimal on a bicycle.

To calculate the applied force at the patch area you will need to calculate the geometric center of the displacement of the bicycle and its rider, and you will need to calculate the center of gravity; you will need to know the geometric volume of the bicycle and its rider which is usually changing continually due to the wind and flapping parts dangling from the operator and/or the bicycle equipment, and the weight of the bicycle and its rider at its weight center and you will have to calculate the aerodynamic shape's center of gravity which changes continually because it is alive. The formula for evaluating applied forces will come later in this writing. Calculating the geometric center and the center of gravity of the bicycle and its rider requires both geometry and statics. The geometric center and the center of gravity are calculated independently of one another. They are separate entities of calculations all together. Once you have calculated the geometric center you will have calculated the geometric volume of the aerodynamic shape simultaneously.

Calculating the center of gravity requires geometry and statics and once you have calculated the center of gravity you will have to calculate the statics to find the weight center since the actual weight center and the center of gravity of an actual bicycle and its rider will probably not coincide in most respects. This leaves the question "How do I calculate the center of gravity?" since the weight center is different from it. Calculating the center of gravity is shown in the drawings on the up-coming pages, it requires the same technique as calculating the geometric volume of the aerodynamic shape to find the geometric center although this time you do not find a center you simply calculate two volumes that are equal (two times) and apply their intersection of two planes.

These concepts will align vector forces in equal and opposite reaction between the wind acting on the bicycle and its rider and the patch area between the pavement and the tire: however, the bicycle and its rider will not behave adequately under high stress aerodynamic conditions because the weight center does not coincide with the calculated center of gravity. This erroneousness between the weight center and the calculated center of gravity is not an error; the weight center simply must be counterbalanced to align the weight center to the center of gravity. The only problem with that is now that the bicycle and its rider is aerodynamically balanced it will probably fly unless the vector between the center of gravity (now the weight center congruent to the center of gravity (BB')) and the aerodynamic center points anywhere else around the center of gravity besides up and back (in the left side view), and going fast may become a problem. However, when the bicycle and its rider are unbalanced like the vector pointing anywhere else problem, the bicycle and its rider are unstable and in an emergency one can lose control and this is probably caused by the aerodynamics of an unbalanced bicycle and its rider.

Balancing aerodynamically to create "ground-hugging" would seem to be the best way to aerodynamically construct the aerodynamic geometry of any aerodynamic shape that is designed to move across the ground. Aerodynamic "ground-hugging" may be constructed in to the aerodynamic shape by aligning the vector (center of gravity, aerodynamic center (left side view)) to be down and back with the center of gravity at the top left, and the geometric center of the aerodynamic shape towards the bottom right. This is sort of a mirror of aerodynamic lift and is aerodynamically stable. The sum of the moments may be found equal to zero at the center of the aerodynamic shape with the applied forces of the patch area and BB' at their respective placed distances from the center of the aerodynamic shape. This balancing at the geometric center of the aerodynamic shape is contrary to flight characteristics the sum of the moments equal to zero is not balanced at the geometric center of the aerodynamic shape for a flying machine, and is for a ground-hugging machine. The vector should be down and to the right from BB' in the left side view, the center of the aerodynamic shape being to the bottom right from BB'. Finding the harmony between the volume of the aerodynamic shape and the weight at the center of gravity and aligning the vector angle to apply the nominal force with the minimum of applied power at the optimum velocity to sustain lift requires some evolution, whether you want the lift to go up or down.

Now there is the act of flying a bicycle which is what this entire manuscript is all about. The science fiction of this prospect is doing the calculations and actually engineering a proper bicycle powerplant that will propel a human powered aircraft in to space, not just the surrounding space, where you have room, or into the air…but space, outer-space where there is no air, to achieve an altitude of more than 90 miles high. However that presents a problem because the human powered spacecraft is only propelled by air (don't get any hair-brained schemes that a human powered spacecraft can be propelled by liquid fuel, hypergolic fuel, or solid fuel rockets) and once the air is gone there is nothing to propel the spacecraft any longer, so it will simply fall back in to the atmosphere (there is a prospect of orbital rate combinations to sustain altitude but I don't think any human powered spacecraft will be able to sustain orbit at this time without fuel propelled assistance, since the speed to achieve orbit is considerable and cannot be sustained in the atmosphere): However, getting that much force out of any human powered powerplant design is still a prospect of discovery.

Anyway, the delivery system of the powerplant turbines must maintain the optimum output performance of applied force at the required limits of tolerances to sustain ascent while running out of air density with increasing altitude. This means that the turbine must continually accelerate as air density decreases while maintaining the aligned proper performance output delivery of the thrust on continuous ascent. The negligibility of this prospect is that once the tolerance of the state of flight has been achieved, then being able to sustain a constant cadence while increasing the output of the turbine with decreasing air density simultaneously only applies the required powerplant to be able to sustain lift of flight from the ground and being able to have a range of cadence through the powerplant's power curve throughout the event, as long as the air density is decreasing the turbine will continue to accelerate. A powerplant capable of doing this job may result in the incapacity of the operator to obtain a cadence at lift on take-off. The following concepts are theoretical: It may be necessary to implement a secondary powerplant calculated to succeed to an altitude where it may be jettisoned only to be taken over by another powerplant which powerplant is calculated to engage at an altitude where it will operate through the range of the operator's cadence and succeed to a higher altitude. This presents the problem of having too many stages with respect to which the operator may not be able to obtain normal lift speed at take-off because of their numbers being too heavy overall. A human powered aircraft capable of reaching greater than 90 miles altitude may have to be launched from a mother-ship to reduce the weight of having too many powerplants to jettison after take-off. Also, jettisoning powerplants will alter the weight distribution of the human powered spacecraft and the wings will have to be designed to move to compensate for the weight changes.

Still, the turbine can only move air up to the speed of sound, exceeding the speed of sound with a turbine can only be done if the turbine blade vanes displacement is bigger than the spacecraft. Once the air "breaks" from the turbine vanes then the turbine vanes do not develop any thrust in excess of the speed of sound. However, as air density decreases, the sound barrier becomes easier to break. This then presents the problem of turning a turbine that only pumps air where there is low density air. More turbine vanes are needed to pump the same amount of air as air density decreases with increasing altitude.

This theory is the more practical theory: Slow the spacecraft down by increasing its lift capability, align its flight characteristics to air density at 90 miles altitude for its weight (if in fact there is air at 90 miles altitude), apply the most powerful powerplant that this analytical mathematical function is capable of, and build a spacecraft that is so slow on take-off that it will reach optimum or maximum speed with respect to its design parameters above 90 miles altitude. The trick is to design the spacecraft at 90+ miles high "air" density on the ground. Knowing the "air" density at 90 miles altitude is simple enough (if in fact there is "air" at 90+ miles high altitude), although the author won't go into designing an aircraft capable of reaching 90+ miles altitude here.

Constructing a powerplant that will sustain a normal take-off speed on the ground and fly to 90+ miles altitude may result in a very slow cadence at ground level on take-off which can be compensated for by a bypass system which will allow the operator to crank faster while maintaining the normal take-off thrust which would otherwise be a very slow cadence. The bypass system may be gradually closed with increasing altitude as air density decreases allowing the operator to maintain a constant cadence while the turbine accelerates in to less dense air on ascent.

This eliminates the jettisoning powerplants problem and allows there to be only one powerplant. However, the quantity of turbine vanes will be an issue although if the spacecraft is designed at 90+ miles altitude "air" density then the turbine will be designed at 90+ miles altitude and air density and the problems should solve out as a result. Once the spacecraft is built and flies successfully then there will be no question of putting the first bicycle in space remaining.

Chapter 2

The Equation

Balancing an aircraft and designing a human powered helicopter has turned out to be a 24+ years research project on the part of the author. There will be no inventive steps in the development of the final equation shown here in the following figures. On the following pages are drawings and writing, calculations with solutions but the problems are not in numerical form. There problems are in the science of mechanics and the equations are derived from actual values for variables. The drawings and writing are both freehand and computer aided design drawings and writing. Explained in these drawings and writing is the final development of calculating the engine performance, and thrust of a typical aircraft. Also the vector, which is typical for all aircraft, which is the hypotenuse of the lift and drag is finally solved in its discovery using it for all calculations. The final vector found is simple and elementary. It is the vector between the center of gravity (for which the weight center is counterbalanced coincidental to the center of gravity) and the geometric center of aerodynamic shape (the geometric center of the aircraft body geometry), and is placed at the center of gravity plane parallel to the trajectory side view of the aircraft in most cases, the vector being up and to the right in the left side view and passes through the aerodynamic center. Two formulas describe the thrust. The formulas are different but when the values for the variables are applied the solutions are the same. One formula is basic trigonometry; the other formula is the fluid mechanics of inertia formula multiplied to some standard trigonometric ratios the same as in the other formula. Values for variables are in American Standard Units: cubic inches, pounds per cubic foot, and pounds. When concluding the calculations you will have air density, displacement of the aerodynamic shape in cubic inches, location of the center of gravity, the geometric center of the aerodynamic shape, the final vector for calculations, the weight of the aircraft, thrust force in pounds, engine performance in pounds, and a designed model aircraft. The next five or six pages are the drawings and writing of this prospect. Keep in mind, the calculator exponent function does not give the right answer when calculating a decimal point exponent.

Base formula

$$(A^p{}_{Nf})$$

0   A^1 A^2 A^3 A^n $\dfrac{B}{Nf}$ A^{n+1}

$$Nf\,A^{\left(\frac{\frac{B}{Nf} - A^n}{A^{n+1} - A^n}\right) + n} = A^p\,Nf = B$$

$$\boxed{A^p\,Nf \neq \frac{B}{Nf} \quad \text{when } Nf \neq 1.}$$

Solve for p.

n = an integer whole number.

N = density

f = volume

B = Force, lbs.

A = base , 2.

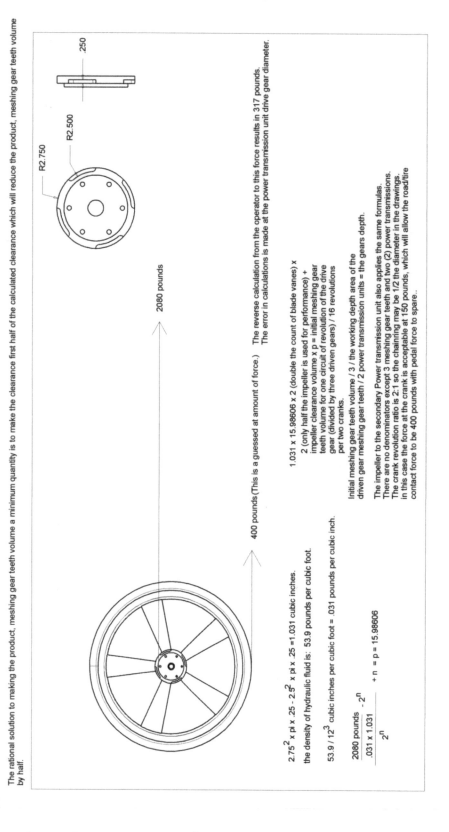

The rational solution to making the product, meshing gear teeth volume a minimum quantity is to make the clearance first half of the calculated clearance which will reduce the product, meshing gear teeth volume by half.

R2.750

R2.500

.250

2080 pounds

400 pounds (This is a guessed at amount of force.)   The reverse calculation from the operator to this force results in 317 pounds.
The error in calculations is made at the power transmission unit drive gear diameter.

2.75^2 x pi x .25 - 2.5^2 x pi x .25 = 1.031 cubic inches.

the density of hydraulic fluid is: 53.9 pounds per cubic foot.

53.9 / 12^3 cubic inches per cubic foot = .031 pounds per cubic inch.

$$\frac{2080 \text{ pounds}}{.031 \times 1.031} - \frac{2^n}{2^n} + n = p = 15.98606$$

1.031 x 15.98606 x 2 (double the count of blade vanes) x
2 (only half the impeller is used for performance) +
impeller clearance volume x p = initial meshing gear
teeth volume for one circuit of revolution of the drive
gear (divided by three driven gears) / 16 revolutions
per two cranks.

Initial meshing gear teeth volume / 3 / the working depth area of the
driven gear meshing gear teeth / 2 power transmission units = the gears depth.

The impeller to the secondary Power transmission unit also applies the same formulas.
There are no denominators except 3 meshing gear teeth and two (2) power transmissions.
The crank revolution ratio is 2:1 so the chainring may be 1/2 the diameter in the drawings.
in this case the force at the crank is acceptable at 150 pounds, which will allow the road/tire
contact force to be 400 pounds with pedal force to spare...

Initial patch area torque force ratio

$(lbs) BB' = h$

$h \sin \alpha + h \cos \alpha = thrust \, (lbs)$

$CD \, 2^{\wedge} \left(\dfrac{\frac{BB'}{CD} - 2^n}{2^{n+1} - 2^n} + n \right) \sin \alpha \, +$

$CD \, 2^{\wedge} \left(\dfrac{\frac{BB'}{CD} - 2^n}{2^{n+1} - 2^n} + n \right) \cos \alpha = thrust \text{ of vessel body volume.}$

the necessity to have to calculate all the vectors for all the subsegments is negligible; all you will have to evaluate is the final result from the geometry of the entire vessel body BB' to AA' with respect to the angle α resulting from the equal volumes opposite perpendicular at AA' and BB' (solid body geometry).

Multiply by $\dfrac{\text{thrust}}{\text{Vessel body volume}} \left(\dfrac{\frac{\text{Vessel body volume}}{\text{blade vanes volume}} - 2^n}{2^{n+1} - 2^n} + n \right) \times \text{blade vane volume}$

$C \cdot bvv \, 2^{\left(\dfrac{\frac{\text{thrust lbs}}{C \cdot \text{blade vanes volume}} - 2^n}{2^{n+1} - 2^n} + n \right)} \cos \beta + C \cdot h \, bvv \, 2^{\left(\dfrac{\frac{\text{thrust lbs}}{C \cdot bvv} - 2^n}{2^{n+1} - 2^n} + n \right)} \sin \beta$

$= $ applied engine performance force at blade vane geometric center.

Propeller blade vanes volume

$C \cdot bvv \, 2^{\left(\dfrac{\frac{BB'}{(C \cdot bvv / 12^3)} - 2^n}{2^{n+1} - 2^n} + n \right)} \cos \alpha \cos \beta + C \, bvv \, 2^{\left(\dfrac{\frac{BB'}{(C \cdot bvv / 12^3)} - 2^n}{2^{n+1} - 2^n} + n \right)} \sin \alpha \sin \beta = thrust.$

$C = $ air density, bvv in cubic inches
BB' is in pounds.
Thrust is in pounds

Final vector, rough draft

$(lbs.)\ BB' = h$

$h\ \text{sine } w + h\cos w = \text{thrust (lbs.)}$

$$CD2^\wedge \left(\frac{(CD/12^3)\ \dfrac{BB'}{2^{n+1}-2^n} - 2^n}{} \right) + n)\ \text{sine } w\ +$$

$$CD2^\wedge \left(\frac{(CD/12^3)\ \dfrac{BB'}{2^{n+1}-2^n} - 2^n}{} + n \right)\ \cos w = \text{thrust (lbs.) of vessel body volume.}$$

The necessity to have to calculate all the vectors for all the subsegments is negligible; all you will have to evaluate is the final result from the geometry of the entire vessel body BB' to AA' with respect to the angle w resulting from the equal volumes opposite perpendicular at AA' and BB' (Solid body geometry).

$$Cxbvv\ 2^\wedge \left(\frac{(C\ x\ bvv / 12^3)\ \dfrac{BB'}{2^{n+1}-2^n} - 2^n}{} + n \right)\ \cos w \cos v\ +$$

Final vector and calculations

$$Cxbvv\ 2^{\wedge}\left(\cfrac{BB' - 2^{n}}{\cfrac{(C \times bvv / 12^{3})}{2^{n+1} - 2^{n}}} + n\right)\ \text{sine } w\ \text{sine } v = \text{thrust}$$

bvv = blade vanes volume of turbine blade or propeller in cubic inches

C = air density in pounds per cubic foot

BB' is in pounds

Thrust is in pounds

Sine v is the angle of association made with the first thrust of vessel body volume made with angle w applied to the blade vanes hypotenuse and its adjacent side. The original thrust of vessel body volume applies to the hypotenuse of the blade vanes applied load of what would be WR (wind resistance) bearing on the aircraft at h sine w + h cos w = thrust (lbs.).

Remember that the calculator will not calculate the base exponent correctly in accordance with the rules of fluid mechanics of inertia.

Final vector, completing calculations

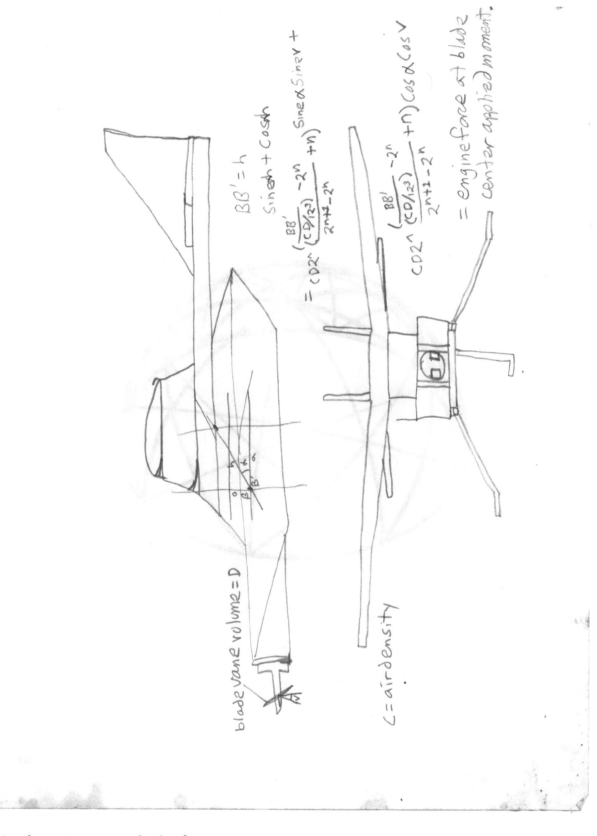

blade vane volume = D

BB' = h

Sine h + Cos h

$$= CD2^n \left(\frac{\frac{BB'}{(CD/123)} - 2^n}{2^{n+} - 2^n} + n \right) Sine \alpha Sine \gamma +$$

C = air density

$$CD2^n \left(\frac{\frac{BB'}{(CD/123)} - 2^n}{2^{n+1} - 2^n} + n \right) Cos \alpha Cos \gamma$$

= engine force at blade

center applied moment.

Final vector, rough draft

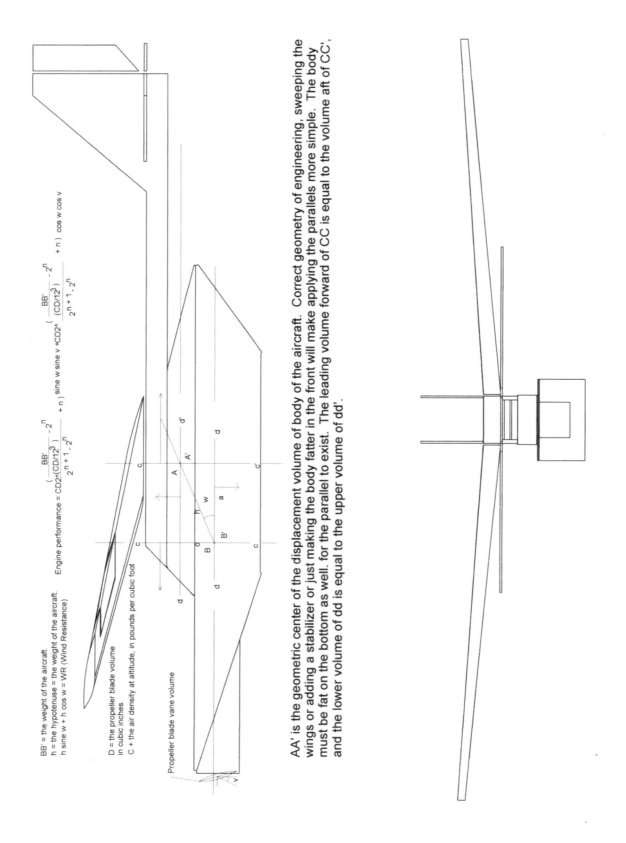

BB' = the weight of the aircraft.
h = the hypotenuse = the weight of the aircraft.
h sine w + h cos w = WR (Wind Resistance)

Engine performance = CD2^(CD/12³) $(\frac{\frac{BB'}{(CD/12^3)} - 2^n}{2^{n+1} - 2^n} + n)$ sine w sine v +CD2^ $(\frac{\frac{BB'}{(CD/12^3)} - 2^n}{2^{n+1} - 2^n} + n)$ cos w cos v

D = the propeller blade volume in cubic inches

C + the air density at altitude, in pounds per cubic foot

Propeller blade vane volume

AA' is the geometric center of the displacement volume of body of the aircraft. Correct geometry of engineering, sweeping the wings or adding a stabilizer or just making the body fatter in the front will make applying the parallels more simple. The body must be fat on the bottom as well. for the parallel to exist. The leading volume forward of CC is equal to the volume aft of CC', and the lower volume of dd is equal to the upper volume of dd'.

Final vector, CAD drawing

Then there's the impact reaction force(s) reacting on both BB' and AA' at velocity in equal and opposite reaction to impact 100% elastic when applied to F, and $h\sin2\alpha + h\cos\alpha = $ Thrust = F. Knowing the velocity where BB'=1 impact reaction force, as well where AA'=1 impact reaction force has not yet been identified, common items are not yet made.

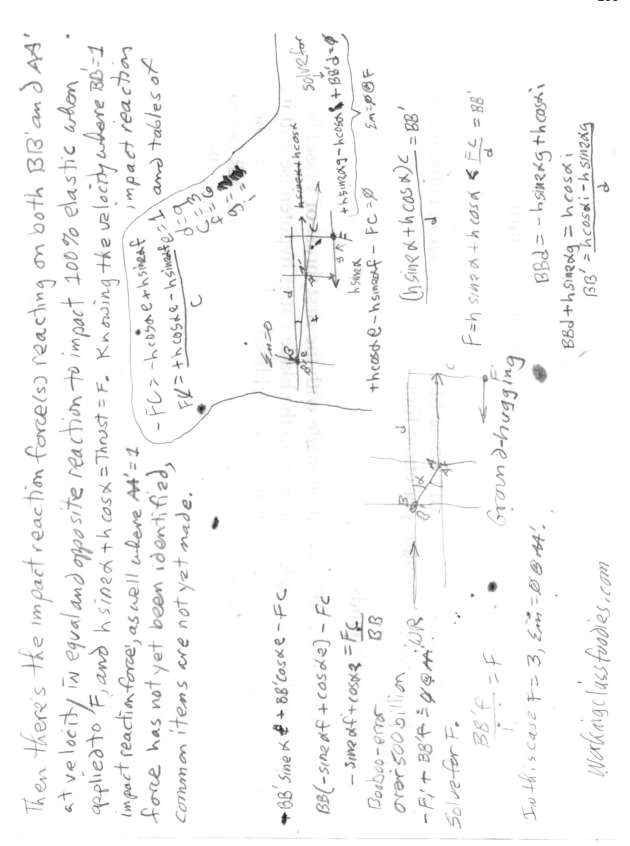

$-FC = -h\cos\alpha\theta + h\sin\alpha f$

$FC = +h\cos\alpha\theta - h\sin\alpha f$

$\Sigma m = 0$ $+h\sin2\alpha\theta + h\cos\alpha$

$+h\sin2\alpha g - h\cos\alpha f + BB'd = \emptyset$

$+h\cos\alpha f - h\sin\alpha f - FC = \emptyset$ $\Sigma m = \emptyset F$

solve for

$\dfrac{(h\sin\alpha + h\cos\alpha)c}{d} = BB'$

$F = h\sin2\alpha + h\cos\alpha \dfrac{FC}{d} = BB'$

$BBd = -h\sin2\alpha g + h\cos\alpha i$

$BBd + h\sin2\alpha g = h\cos\alpha i$

$BB' = \dfrac{h\cos\alpha i - h\sin2\alpha g}{d}$

$+ BB'\sin\alpha \emptyset + BB'\cos\alpha - FC$

$BB(-\sin\alpha f + \cos\alpha 2) - FC$

$-\sin\alpha f + \cos\alpha 2 = \dfrac{FC}{BB}$

Booboo-error
over 500 billion

$-F_i + BB'F = \emptyset\emptyset m i.WR$

Solve for F.

$\dfrac{BB'F}{i} = F$

In this case F = 3, $\Sigma m = \emptyset @ AA'$.

Ground-hugging

Workingclassfoodies.com

Ground hugging, rough draft

Chapter 3

Balancing An Airplane

Balancing an airplane is relatively simple once you have calculated the aerodynamic center of the aerodynamic shape. Although compound complex curved aerodynamic shapes make this part of the calculating process very hard it is more simple to make aerodynamic shapes from rectangular boxes, prisms, hemispheres, and pyramids or other elementary geometric shapes which can be strung together in to an aerodynamic shape which is more easily calculated for these problems: bisecting the aerodynamic shape at the geometric center with a plane perpendicular to the direction of travel (the trajectory) and applying a parallel plane forward of the first plane until the volume of the aerodynamic shape ahead of the second plane is equal to the volume trailing the first perpendicular plane, will give you the first line of intersection viewed from the left side view of the aerodynamic shape. Next, a plane made through the aerodynamic center of the aerodynamic shape parallel to the trajectory moves a parallel plane below it until the volume below the parallel plane made second is equal to the volume above the plane made first through the aerodynamic center of the aerodynamic shape. The intersection of each second plane of both first planes is the calculated center of gravity. The resulting vector between the intersection of the two second planes and the intersection of the two first planes is the vector which is the final vector (angle w) used to calculate the thrust knowing the turbine blade vanes volume and their specific resulting final vector (angle v) with respect to which the Sines and Cosines for both the aerodynamic shape of the aircraft body displacement and the turbine blade vanes are multiplied ((coefficient) x sine w x sine v + (coefficient) x cos w x cos v) to find the engine performance. The coefficient is the relationship between the weight of the aircraft (BB' in pounds), the air density at altitude (C in pounds per cubic foot), and the displacement of the turbine blade vanes (D in cubic inches), and cancelling cubic inches and pounds/cubic foot with cubic inches/cubic foot) and the formula for Fluid Mechanics Of Inertia: $CD2^{\wedge}(BB'/(CD/12^{\wedge}3) - 2^{\wedge}n/2^{\wedge}n+1 - 2^{\wedge}n +n)$ x sine w sine v + $CD2^{\wedge}(BB'/(CD/12^{\wedge}3) - 2^{\wedge}n/2^{\wedge}n+1 - 2^{\wedge}n + n)$ x cos w cos v = engine performance force in pounds. This formula is relatively simple, it requires adding, subtracting, multiplying, and dividing. It is simply a complex fraction in the exponent where BB' is the weight of the aircraft in pounds, C is the air density in pounds per cubic foot, D is the volume of the air craft turbine blade vanes in cubic inches, and all the units cancel and the result is a dimensionless number in the fraction in the exponent. Quantities in parenthesis are worked first, or together as quantities, it is difficult to understand in this example what is above the divisor and what is in line with it. CD2 and + n) are in line. The divisor in BB' / $(CD/12^{\wedge}3)$ is in line with $- 2^{\wedge}n$. You can see the equation in the previous pages of drawing and writing. The equation is relatively simple. $2^{\wedge}n$ is just the next whole number lower than what BB'/$(CD/12^{\wedge}3)$ is equal to: e.g. 1589 is the quotient: then, $2^{\wedge}10$ is = $2^{\wedge}n$ because $2^{\wedge}11$ is 2048 which is > 1589 and the fraction in the exponent of the base two will not solve. n is a whole number integer.

These are the equations and calculations of the previous five pages of drawings and writing. Their application is Universal and applies to every creation of the Almighty that moves through a medium and sustains altitude or glides, operates, or simply has force and motion whether it's in equal and opposite reaction to a medium or to itself in space where its impact reaction force sustains it on its trajectory in harmony with attractive gravitational forces in the Universe, or Galaxy to simplify things. By the way, gravity doesn't exist in space, only in matter. Keep in mind the vectors in compression point away from one another, and vectors in tension point towards one another. This is contrary to what is taught in statics courses in college. The reason for these vectors directions is when (1) in compression when the beam is elastic the forces are on the ends of the beam, the vectors show resistance to the forces and their arrowheads face away from one another. (2) In tension the same is true for elastic beams: when the forces are in tension the beam will hold together and the vector forces arrowheads face one another in between the applied tensile forces at the beams ends. If the opposite is true the beams are destroyed.

Thus, helicopters action and reaction forces are simultaneous and in Sum of the Moments equal to zero. The helicopter does not move during static alignment of forces although it is in continuous performance of motion in reality. The helicopter applied moments limits of tolerances are 100% elastic in Fluid Mechanics of Inertia calculations. There is no flex in the helicopter during static alignment of forces but for calculating motion of flex force(s) simultaneously in performance. The helicopter (as well the rocket and the jet airplane, submarine, ship, et cetera) is fixed in time during calculations (zero time present) although it is still allowed to be alive in its perpetual motion since it is a mechanism that operates in space and time and can be manipulated to perform activities of required motion continually while it is performing its output alignment of static limits of tolerances for which "zero time present" is always in alignment with respect to the helicopter's output power, even though the helicopter is constantly changing its curve of performance throughout its life during its flights. As well, all these performance variables apply to birds.

Applying the altitude air density at which the powerplant will apply its optimum output limit and aligning the lift blade vanes displacement and blade vanes angle elevation attitude to sustain maximum altitude for the optimum weight of the aircraft (helicopter), which is virtually impossible to control (the author will write about that later), the helicopter should fly at its best when parameters of power curve maximum, blade vanes volume, output revolutions at optimum power limits of tolerances, weight of the helicopter and any other limits of tolerances which may enhance the resulting performance of the helicopter will manage to provide for a reasonable helicopter design throughout its powertrain.

Now this is the cop-out: most of the previous rationale is for consumable propellant fuel powered helicopters. A human powered helicopter requires the static elastic balancing limits of tolerances with respect to the sum of the moments equals zero. Finding where the torque was was what the Revelation into this human powered helicopter was all about. A torque lever applied can affect a depth of meshing gears which gears depth can be moderate if not extreme since the torque lever and the gears radius does not change, the torque remains the same. Meshing gear teeth fill with fluid and become a pump and then torque applies and the gears depth makes the torque apply to drive a hydraulic pump of epicyclical gears, the force of torque is constant despite the depth of the meshing gears teeth. A chain drive sprocket drives the center gear in a planetary gears pump. The center gear is 1/3 the ratio of the 1 sprocket radius, or vice versa, and can be a different ratio. As well the hydraulic fluid coming from the planetary gears pump turns an impeller with a radius of, say, 45 inches or so which impeller having its hydraulic fluid applied to its blade vanes in suction applies torque to the center gear of another epicyclical gear pump which hydraulic fluid turns an impeller which turns the lift blade.

As well, another set of planetary gears powered by the operator's chain drive assembly powers another impeller which drives another planetary gears pump which turns an impeller which drives the tail rotor blade.

Aligning the impellers radii so the calculations for applied forces throughout the entire powertrain are within tolerances of the operator's ability to apply force to the pedals, align the applied force at the pedal by varying the impellers radii with respect to the operator's capability to sustain the applied force at the pedals. These calculations of alignment of the impellers radii simply require ratio and proportion to get the optimum impeller's radius on the second try. The desired operator's force divided into the actual operator's force calculated x the impeller radius = the new impeller radius which will apply the desired operator's force. The lift blade should be 2/3 of the desired load while the tail rotor force should be 1/3 of the desired applied operator's force. The applied force at the pedals and to the operator should be low enough so that the operator can maintain flight for a reasonable period of time. Human powered helicopters are custom designed with respect to the buyer's specifications: inseam, height, weight, waistline, shoe size, nominal strength.

The conditions on a human powered helicopter apply the same as a combustible fuel powered helicopter; the lift performance and tail rotor performance must remain at flight performance output power throughout the duration of flight. One cannot "coast" on a human powered helicopter without immediately losing lift performance and tail rotor control.

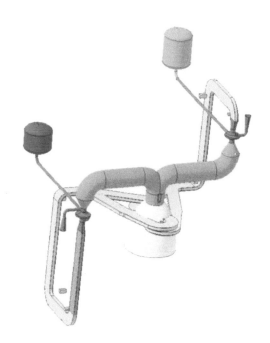

Rocket Motor Sketch, CAD drawing

Maylene and the Sons of Disaster
Rocket motor equations.

Whistle
Multi-tool
Compass

$$= \frac{2^{21} \text{ pounds rocket exhaust force}}{\text{internal area of the rocket nozzle to 3in.}} \times \frac{\text{Port area or conduit area}}{\text{internal area of the rocket nozzle 2in.}}$$

$$= \text{applied force to port area or conduit area (B). lbs.}$$

$$\frac{\left(\frac{\text{Exhaust force lbs } 2^{21}}{\text{internal volume of the rocket nozzle 3ft}} \times \text{displacement of turbine gallery 3in} \right) - \frac{2^n}{2^n} + n = \frac{r}{2} \text{(denom.)}}{2}$$

$$\frac{B}{\text{(denom)} \; 2} \qquad \frac{123 \; 3in/3ft}{2'} \qquad 2$$

• optimum
= applied force to turbine gallery, lbs.

To apply throttle it must be possible to reduce the density of the exhaust force in the conduit to the turbines by allowing less exhaust in the conduit volume thereby reducing the exhaust's density by causing it to expand in the conduit volume. The throttle should be closed and the rocket motor idle applied output force should sustain or be equal to the weight of the rocket plus its occupants.

First multiplier
= applied force to the turbine gallery

$$\left(\frac{\text{Exhaust force lbs } 2^{21}}{\text{internal volume of rocket nozzle 3ft}} \times \frac{\text{Throttle closed area}}{\text{Throttle full open area}} \times \frac{\text{Turbine blade 2}}{\text{vanes volume 3in}} \right) - \frac{2^n}{2^n} + n \mathbf{)} \times \frac{\text{desired turbine revolutions}}{\text{to sustain port area}}$$

e/o reaction force
$$\times \frac{\text{volume of turbine blade vanes}}{\text{swept vol. of turbine vanes}}$$
turbine vanes vol.

$$123 \; 3in/3ft$$

269

+ clearance volume in the turbine gallery x (first multiplier) = applied static force to the turbine gallery to sustain idle.

Now this is the problem: what is the velocity of a volume of air when its simultaneous equal and opposite reaction force at impact is equal to its weight?

It would seem to hold true that different densities have different velocities at which their impact reaction force is equal to their weight.

Rocket Motor Equations Completed, rough draft

Rocket motor equations:

$$\frac{2^{21} \text{ pounds rocket exhaust force x port area or concuit area}}{\text{internal area of the rocket nozzle, }{}^2\text{in.}}$$

= applied force to port area or conduit area (B), lbs.

$$\text{B}$$

$$(\text{denom.})\ 2\ \frac{\left(\text{Exhaust force, lbs, } 2^{21}\right.}{\substack{\text{internal volume of the}\\ \text{rocket nozzle, } {}^3\text{ft.}}}\ \text{x displacement of turbine gallery, } {}^3\text{in.}$$

$$12^3\ {}^3\text{in} / {}^3\text{ft.}$$

$$\frac{\dfrac{2^n - 2^n}{2^1} + n = 2^r\ \text{x denom.}}{2^1} = \text{optimum applied force to the turbine gallery, lbs.}$$

Rocket Motor Equations, CAD program

To apply throttle it must be possible to reduce the density of the exhaust force in the conduit to the turbines by allowing less exhaust in the conduit volume thereby reducing the exhaust's density by causing the gas to expand in the conduit volume. The throttle should be closed and the rocket motor idle applied output force should sustain or be equal to the weight of the rocket plus its occupants.

First Multiplier

applied force to the turbine gallery

$$\left(\frac{\text{exhaust force in pounds, } 2^{21}}{\text{internal volume of the rocket nozzle, cubic feet.}} \right) \times \frac{\text{throttle closed area}}{\text{throttle full open area}} \times \frac{\text{turbine blade vanes volume, cubic inches}}{}$$

$$-\frac{2^{n}}{2^{n}} + n \Big) \times \text{desired turbine revolutions to sustain port area equal and opposite reaction force}$$

$$\times \text{volume of turbine blade vanes} \times \frac{\text{swept volume of turbine vanes}}{\text{turbine vanes volume}}$$

$$+ \text{clearance volume in the turbine gallery} \times (\text{first multiplier})$$

Rocket Motor Equations, Cad program

= applied cubic volume to the turbine gallery to sustain idle.

Now this is the problem: what is the velocity of a volume of air when its simultaneous equal and opposite reaction force at impact is equal to its weight?

It would seem to hold true that different densities have different velocities at which their impact reaction force is equal to their weight.

Rocket Motor Equations, CAD program and remaining problem

Chapter 4

The Rocket Motor

The previous pages show a mirrored parts attempt to show a rocket motor in its most basic simplicity and some of the mathematics used in Fluid Mechanics of Inertia to design a rocket motor. None of the variables in the assembly drawing are calculated for precision tolerances. The assembly drawing is simply pieced together from scratch and no parts in their specific tolerances align or match with respect to the equations necessary to make a rocket motor work. The following description will be an attempt to describe the mathematics involved in the assembly drawing of the elementarily simple rocket motor on the previous pages.

A rocket motor's combustion must be aligned in performance at static alignment throughout the powerplant. The turbines in the turbochargers are aligned with static alignment, revolutions, and clearance tolerances so the elementary equation for that prospect of idea is relatively simple and will be described in equations. These equations will be in "Fluid Mechanics of Inertia's" base formula, so all there will be involved will be adding, subtracting, multiplying, and dividing, so a sixth grader should be able to design this attempt to describe a rocket motor's performance using the drawing and the base formula for Fluid Mechanics of Inertia.

Practice with the base formula for Fluid Mechanics of Inertia will be applied for the duration until the time when human powered helicopters becomes the subject in this manuscript. In the meantime the first determination to exercise the base formula for Fluid Mechanics of Inertia will be by using a rocket motor to apply the values for variables in the formula to describe the output performance of this particular rocket motor concept in operation. This rocket motor design would seem to be self perpetuating in concept: however, there are certain parameters inside the rocket nozzle which must be aligned for this rocket motor to work. This rocket motor needs combustible propellant and oxidizer to operate.

The fuel and oxidizer inlet ports at the rocket nozzle have to be a certain area which is proportional to the area of the conduit leading to the turbines. The combustion of the fuel and oxidizer apply force to the fuel and oxidizer inlet areas by applying torque force at the turbines to the fuel and oxidizer. The combustion of the fuel and oxidizer also has a standing force at the conduit area at the rocket nozzle that acts in equal and opposite reaction to the resulting force applied at the turbines to overcome the combustion force at their port areas in the rocket nozzle. The turbines sustain equal and opposite reaction force to the force within the conduit to the turbines at the required revolutions to sustain the force of the fuel and oxidizer at their port areas inside the rocket nozzle. The combustion in the rocket nozzle is sustained and can vary with throttle and all the variables must vary simultaneously to sustain the combustion in the rocket nozzle.

Being able to exceed the static force at the port areas for the fuel and oxidizer to the rocket nozzle allows for the rocket motor to increase combustion and raise the force applied inside the rocket nozzle allowing for acceleration of the equal and opposite reaction performance. What exactly these limits of tolerances are would require testing and research to find the exact match for all of the applying variables: rocket nozzle shape, fuel and oxidizer inlet port areas, conduit area to the turbines, turbines applied force to the fuel and oxidizer with leeway for acceleration forces in the turbines output capability, and simultaneous alignment of all applicable variables in performance being accurate.

Since fuels and oxidizers have various ratios to combine them for complete combustion the associated areas will also align along with their particular turbines specifications to sustain all the required limits of tolerances of complete combustion in the rocket nozzle while allowing for acceleration with respect to the turbine's capability to increase output force to the fuel and oxidizer flow to the rocket nozzle under reaction force of combustion against the inlet port areas. In the assembly drawing on the page of the elementary rocket motor concept the components are exaggerated in most respects.

It may not take as much force as one may expect to sustain combustion in a rocket nozzle. Flooding the rocket nozzle with the fuel and oxidizer and igniting it, or having it ignite if it is a hypergolic fuel and oxidizer combination, and continuing to allow the fuel and oxidizer to flow manually, e.g. with starter motors and a battery turning the turbines, will deliver force to the conduit which drives the turbines when the combustion begins, successfully with an excess of force while the governor will allow the engine to start sustaining the equal and opposite reaction forces managing a running rocket motor while a governor will allow the forces inside the rocket nozzle to accelerate the alignment of all the variables to the desired performance output of the rocket motor concept to idle. This requires more research to get a rocket motor output performance which will apply its difference of actual fuel/oxidizer combustion force minus the applied equal and opposite reaction forces of the applied turbines through the conduit area at the rocket nozzle measured at the turbines equal and opposite reaction forces in sum, and the resulting difference of applied force is the rocket exhaust thrust. The applied formula for Fluid Mechanics of Inertia describes the limits of tolerances that a rocket may possess for its rocket motor exhaust thrust force to sustain a comparable successful test of a rocket on the first try.

The fluid mechanics of aerodynamics are taken in to consideration: Atmosphere reacts on the spacecraft from the moment it takes off. The spacecraft must sustain equal and opposite reaction force against the atmosphere plus acceleration in to the atmosphere while if the spacecraft is occupied the acceleration of the spacecraft cannot kill the occupants and the capability of the rocket motor thrust output must have a range through which the rocket motor operates which will sustain the aerodynamic force of the atmosphere as the atmosphere becomes less dense with altitude and the rocket accelerates simultaneously which acceleration cannot kill the occupants as gravity becomes less with altitude as well, the difference in gravity can be transformed in to acceleration of the spacecraft, the rocket motor must be able to sustain the maximum aerodynamic force (in pounds) on the spacecraft's displacement (in cubic inches) up to the altitude where its velocity and the aerodynamic force are at their maximum, increasing altitude with increasing velocity will only maintain the same aerodynamic force or the aerodynamic force will diminish if the spacecraft does not accelerate in to less dense air fast enough. (see supersonic calculations in the following drawing and diagrams (next).) I do believe this altitude moment is called "Go At Throttle Up": gravity is diminishing, atmospheric air density is dropping faster than the spacecraft is accelerating and the spacecraft can get altitude by going straight up with nominal thrust as aerodynamic force diminishes, the spacecraft accelerates faster, the occupants experience constant steady thrust (equal and opposite reaction forces), and the rocket motor performs within its range of power curve with the capability of sustaining the velocity of the spacecraft in equilibrium with gravity at orbital altitude.

Some trigonometry is used during orbital entry to balance the apogee and perigee on the orbit of the spacecraft. This leads to the cubit of the Earth (1 unit cubit) to be useful in ascertaining all the variables of interplanetary space travel. Now that man knows how to orbit the Earth with a spacecraft it is possible to use all the variables values of Earth as 1's (units cubit in the denominator) in their relationship with other planets: orbital altitudes, planetary mass, diameter of the planet, anything that can be determined from calculations that are useful for ascertaining planetary ballistics around Earth can be applied to determine values for extraterrestrial bodies provided one can orbit them. Any extraterrestrial body can be orbited, it depends on one's velocity and altitude mostly in comparison to the 1 unit cubit of the respective variable one is using to calculate an orbital trajectory on approach. There are also intervals on approach, how rapidly one is approaching the celestial body. Intervals are coefficient with the altitude and the base two fold exponent. How many altitudes there are between the orbital tangent at altitude and the distance the spacecraft is from the orbital tangent point are proportional with respect to the time it takes to cross the distance and Earth's 1 unit cubit for the same proportional alignment limits of tolerances with respect to the same variables. Planets can be any size and any density, so a planet that's

Earth sized could be more dense (or less dense) so if one entered orbit at the same altitude one would enter orbit at Earth one would have to be going a proportional velocity with respect to the planets densities ratio x the velocity unit cubit of Earth at the same altitude = the extraterrestrial orbital entry velocity, or else the same ratio multiplies to the altitude and the spacecraft enters the orbit at the Earth's coefficient 1 unit cubit velocity, and sustains orbit around the extraterrestrial body. All of Earth's proportional coefficients are equal to 1 until cubit in equations involving interplanetary space travel. Remember the cubit is the denominator quantity and a ratio has a denominator which may be a quantity product or some complex fraction quotient. Using Earth's "cubit" in the denominator and multiplying that denominator to some value of some variable making all the units cancel and the ratio a dimensionless number will result in the coefficient of proportion which can be calculated as a base two exponent which base two and exponent when multiplied to the denominator quantity will equal the original extraterrestrial value. Some consideration for a ratio of intervals can be made but that only complicates the problem.

Gravity doesn't change as acceleration occurs. As altitude applies, the angle thrust makes with tangent to the
orbit defines gravity. The actual thrust vector does not decrease the angle made with the tangent to the orbit
as acceleration velocity reaches y limit at intercept because gravity does not decrease with increasing velocity.

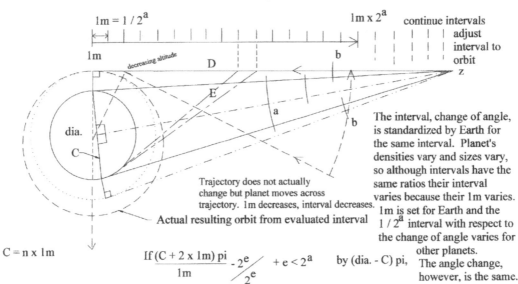

As interval decreases with 1m spacecraft must meet interval time tolerances.

$1m = 1 / 2^a$ $1m \times 2^a$ continue intervals
adjust
interval to
orbit

1m   decreasing altitude   D   b   z

dia.
C
a   b

Trajectory does not actually
change but planet moves across
trajectory. 1m decreases, interval decreases.
Actual resulting orbit from evaluated interval

$C = n \times 1m$

The interval, change of angle,
is standardized by Earth for
the same interval. Planet's
densities vary and sizes vary,
so although intervals have the
same ratios their interval
varies because their 1m varies.
1m is set for Earth and the
$1 / 2^a$ interval with respect to
the change of angle varies for
other planets. The angle change,
however, is the same.

$$\text{If } \frac{(C + 2 \times 1m)\ pi}{1m} - \frac{2^e}{2^e} + e < 2^a \quad \text{by (dia.} - C)\ pi,$$

then D divided into $1 / 2^a$ increments intersects E divided into $1 / 2^a$ increments, and point z may
be nearly determined in 1m increments: $1m = 1 / 2^a$. With respect to (dia - C) pi, the "$< 1 / 2^a$"
increments will have intervals less than the actual $1 / 2^a$ increment interval. Approximating an interval
to displacement time to orbit will require adjustment to the orbit interval displacement time if the
alien planet density is undetermined.

When $1m \times 2^a$ orbit circumference increases by one interval or more, the interval decreases by initial
interval $\times (1 - (2^a - F / 2^a))$, so initial interval aligns 1m on both sides of the planet. F = count of
additional intervals from $1 / 2^a$ goes to $F / 2^a$, F can be any number. Calculation of approach interval
defines limits of tolerances with respect to 1m @ 1m x F apogee theoretically, actual apogee will be
slightly higher. Actual evaluation of $1m \times 1 / 2^a$ interval is theoretical if the density of the alien celestial
body is not known. Theoretically 1m and $1 / 2^a$ align coefficient with respect to 1e where 1m/e and their
$1 / 2^a$ relationships are constant. Alien celestial body densities will cause variations with respect to intervals.

Fly-by-the-seat-of-your-pants planetary orbital entry evaluation

Theory of the Sound Barrier:

air displacement of aircraft volume = V cubic units

the aircraft moves one unit volume and displaces its displacement $\frac{1}{p} = 1v$ cubic volume

N (air density) is specific at altitude.

f (aircraft displacement volume) is constant.

$\frac{B}{p}$ weight of the aircraft

p fold exponent after calculation

$p \times (1v = Vp \frac{1}{p} = V) = \frac{B}{p}$ $\frac{\text{weight of aircraft}}{p}$ = weight of displaced air at altitude.

$\frac{B}{p} < p \times (1v = Vp \frac{1}{p} = V)$

$\frac{\text{weight of aircraft}}{p}$ < weight of displaced air on the ground.

Otherwise, eliminate p in the denominator of B/p, and raise all other p's as base two exponents.

The swept volume becomes $1v2^p = B$.

Theoretical Supersonic Calculations

Chapter 5

A Jet Engine

Once the engine performance for an "airplane" has been calculated knowing the v angle of the turbine blade vanes the combustion force inside the jet engine should be equal to the engine performance force calculated solution. But, the torque from the applied moment of the v angle at the blade vane volume with respect to the lever arm from the v angle to the, in this case, crank journal center multiplied to the engine performance solution should equal the internal combustion force in the combustion chamber of a reciprocating engine. As can be reckoned this combustion force must be sustained continually during the aircraft's flight. This combustion force cannot be sustained by ordinary automobile engines because their connecting rods are not designed to sustain the continuous force greater than the weight of the craft being piloted and sustained at altitude for any length of time. Aircraft internal combustion reciprocating engines have connecting rods specifically designed for sustained engine performance force x the torque ratio = the combustion chamber force.

AVGAS therefore would also be specifically designed to burn cleaner than regular or premium automobile gas at the sustained power required to fly an "airplane". A jet engine performance force is approximately equal to the calculated engine performance force although it must include the additional aerodynamic force at the intake turbines and would double the applied jet engine combustion force in the jet engine combustion chamber. The jet engine pulls and pushes simultaneously. Half of the engine performance force pushes while the other half pulls. The calculated solution with respect to the engine performance is the same as 1 unit cubit. Other calculations can be made using other units cubit: the turbine blade vanes volume, the displacement of the aircraft body et cetera. The jet aircraft should be designed at its maximum altitude with its body displacement volume and its weight being conformable to the engine performance at its optimum output power for the particular air density it will be operating in at the simultaneous altitude, which calculations will lead up to the human powered spacecraft which design parameters are described previously. No drawings are supplied for any jet engines at this time.

Chapter 6

The Extreme Bicycle

Here is a design for a bicycle capable of reaching a speed of 200 miles an hour. One problem is the head-tube needs to be strengthened because of the statics balancing putting most of the weight on the front end. The design parameters of the extreme bicycle are the same as for the human powered helicopter only the output force powers the rear wheel instead of a lift blade. All the calculations are made using the base formula for Fluid Mechanics of Inertia. Calculations are made to include impeller blade vanes volume circuit of revolution (1 u.c. (unit cubit)), clearance volume in the impeller gallery, torque, meshing gear teeth volume circuit of revolution of the center gear, the power transmission clearance volume, applied forces, fold exponents, operator's force at the crank, aerodynamic displacement volume of the extreme bicycle's body displacement volume, applied aerodynamic force at 200 miles an hour (which requires the unsolved problem mentioned above in drawings, for impact force of air at the velocity where it is equal to its weight), and the patch area road/tire force which translates back through the power transmission assembly to the operator at the crank force.

The impellers' diameter is aligned to apply the particular force at the operator which can be sustained at 200 miles an hour. Revolutions of the impeller in performance are multiplied in to the equations multiplied to the impeller blade vanes swept volume (1 u.c.) and the clearance volume in the impeller gallery, which clearance volume x revolutions is added to the applied performance of the impeller at applied revolutions, and the sum of the moments equal zero as long as the operator's force is compatible to the operator for sustained 200 miles an hour.

Calculating the impeller's final flow volume of hydraulic fluid and dividing by the sum area of the center gear circuit of revolution times the number of meshing points of gears in the power transmission and dividing by two power transmissions (x/2) will provide the depth of gears in the secondary power transmission.

The primary power transmission calculations are the same except divided by 4, or 6 depending on how many there are.

The clearance volume in the power transmission is multiplied by the applied load fold exponent to equal the clearance flow volume. Now the clearance flow volume has to be added to the gears depth to sustain its applied force at 200 miles an hour with no hydraulic fluid "slippage". The next page shows the equation for this calculation of making the clearance flow volume add to the gears depth.

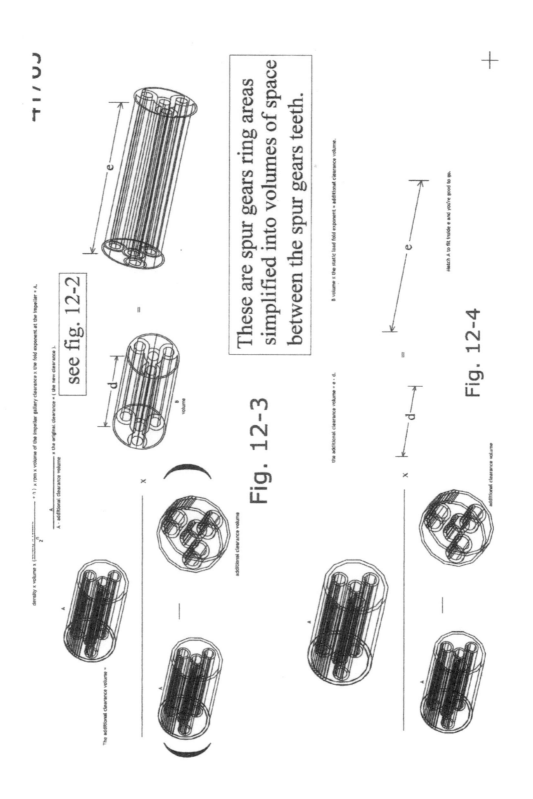

These are spur gears ring areas simplified into volumes of space between the spur gears teeth.

Fig. 12-3

Fig. 12-4

see fig. 12-2

Fluid Clearance Standing Wave Calculations

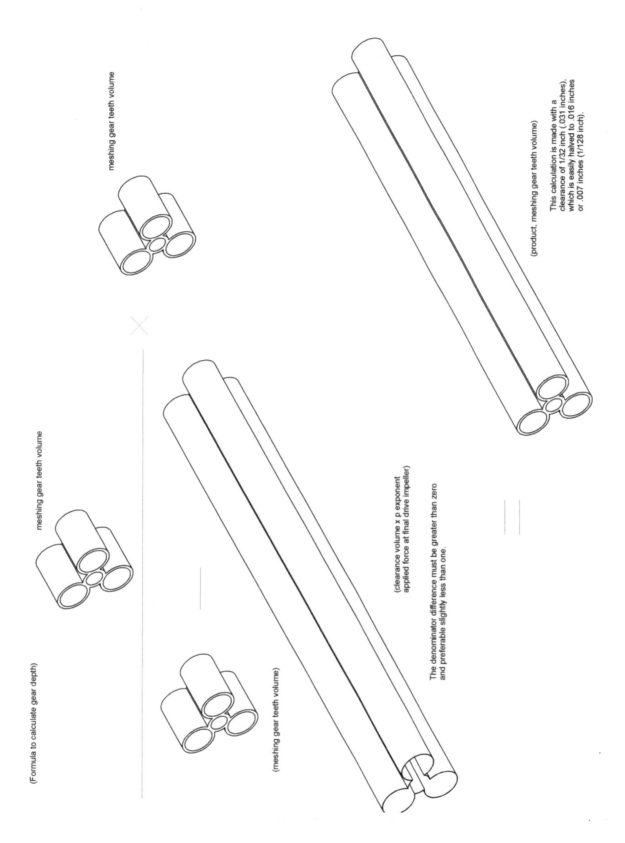

meshing gear teeth volume

(product, meshing gear teeth volume)

This calculation is made with a clearance of 1/32 inch (.031 inches), which is easily halved to .016 inches or .007 inches (1/128 inch).

meshing gear teeth volume

(clearance volume x p exponent applied force at final drive impeller)

The denominator difference must be greater than zero and preferable slightly less than one.

(Formula to calculate gear depth)

(meshing gear teeth volume)

Fluid Clearance Standing Wave Calculations

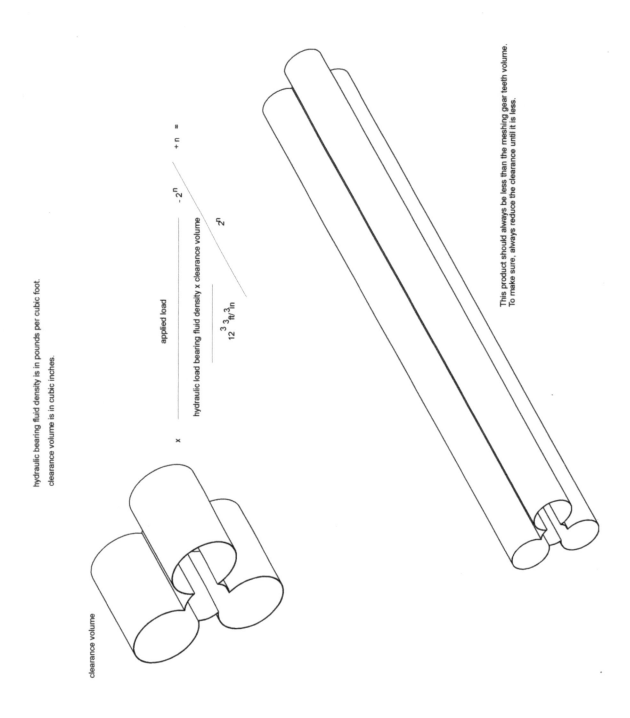

hydraulic bearing fluid density is in pounds per cubic foot.

clearance volume is in cubic inches.

clearance volume

applied load

$$x \quad \frac{\text{hydraulic load bearing fluid density x clearance volume}}{12^3 \, \frac{ft^3 \, in^3}{}} \quad \frac{-2^n}{2^n} \quad + n \quad =$$

This product should always be less than the meshing gear teeth volume. To make sure, always reduce the clearance until it is less.

286

Fluid Clearance Standing Wave Calculations

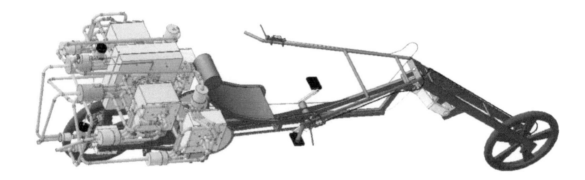

First attempt at 200 mile an hour bicycle (not a failure)

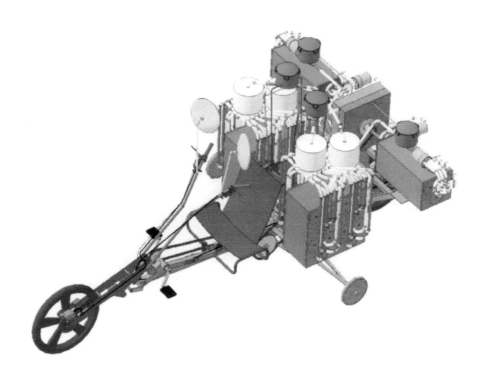

200 mile an hour bicycle design prospect

Once all these equations are satisfied the only remaining alignment of values are the drive train sprockets radii. The sprockets radius on the power transmission for delivering hydraulic fluid to the impeller are three times the radius of the center gear of the epicyclical power transmission. These sprockets are driven by a sprocket of equal radius independent of the power transmissions. On the same axle as the independent sprocket is a sprocket that is ½ the radius of the large sprocket. This small sprocket is driven by a sprocket on the crank that is aligned to be double the radius of that smaller sprocket. Then there is the torque between the crank sprocket and the pedal(s) which finally should provide the operator with a comfortable applied force and comfortable cadence up to 128 revolutions per minute or 2 revolutions per second approximately.

A throttle is provided because if the power train were simply fixed the crank force would be too extreme at the beginning of the run. Hydraulic fluid is allowed to bypass the final output impeller and flow to the primary impeller without flowing through the impeller gallery of the output impeller. This allows for the operator to crank a reasonable cadence while the bicycle is being accelerated from start by closing the throttle. As the throttle is closing more hydraulic fluid is being made to go through the output impeller gallery.

The above paragraphs of chapter 6 are the required points of calculations. Other calculations can be made e.g. the center gears radius of the power transmissions and resulting sprocket radii, the bicycle can be slowed down and recalculated, the bicycle can be designed for children and recalculated et cetera.

$$\frac{\text{Applied pounds}}{\text{metal shear strength}} = r_{in}^2$$

$$2 \times r_1 = \sqrt{\frac{r_1^2 \times pi - r_2^2 \text{ inches}}{pi}} \times 2 = \text{diameter}\ r_1$$

$$2 \times \sqrt{\frac{r_1^2\ in + pi\ r_1^2}{pi}} = r_2^2 \times 2 = \text{diameter}\ r_2$$

(Formulas For Tubing Wall Thickness)

X : N :: X : N   These are applied forces at the chaining.

300 : 48 :: 150 : 24   X : N :: 150 : 24

N = number of teeth in sprocket

N approx. = 36   X = resulting force closest to approximately 220 pounds.
Force varies per occupant.

N / 24 x the circumference of the 24 tooth chaining = the N tooth chainring circumference.

9.9375 in. diameter x pi = 31.22 in. circumference.

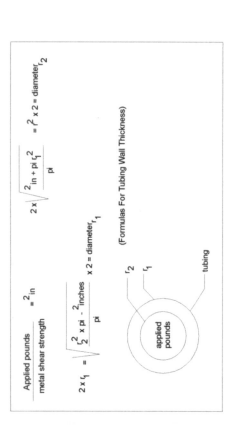

2.625^2 x pi x .25 - 2.25^2 x pi x .25 = .50315 cubic inches x 2 double the blade vanes volume = 1.00629 cubic inches.

$$\frac{2080\ \text{lbs.}}{63.9\ \text{lbs. / cu. ft.}} \times 1.00629\ \text{cu. in.} \quad \frac{2^n + n}{\frac{12^3\ \text{cu. in. / cu. ft.}}{2^n}} = 15.72735 \qquad n = 15$$

(Clearance volume x 15.72735 + 1.00629 x 15.72735 x 60.011649 revolutions per second/ 16 revolutions of the power transmission unit = the secondary meshing gears volume.

The same coefficients apply to the center "box" with respect to the initial meshing gears teeth volume while the center "box" only has 16 revolutions. Torques are readily calculable.

Tubing wall thickness equations

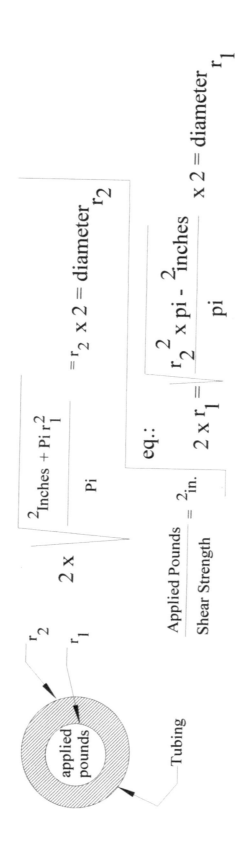

Tubing wall thickness equations

One other calculation has to be made, the tubing wall thickness, since there is considerable suction on the hydraulic fluid in the draw tubing. The head tubing not so much head force if any at all since the head flow just pushes hydraulic fluid up to the reservoirs. Once the hydraulic fluid is in the reservoirs the other sides of the meshing gears teeth in the power transmissions suctions the hydraulic fluid through the impeller gallery from the reservoir around the impeller blades to the suction meshing gears around and to the head side of the power transmissions' meshing gears where it is pumped back to the reservoirs.

As well, there are centrifugal and centripetal forces to contend with for a bicycle during cornering. Inclined curves calculations can be relatively complicated. The following drawings show such complications:

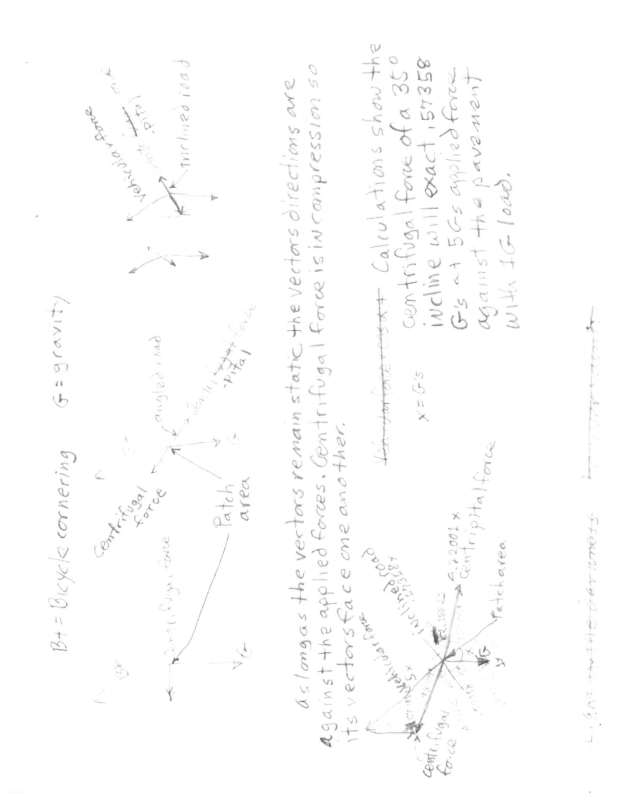

Centrifugal and centripetal force equations of bicycles cornering

Starting behind the driver, the first "boxes" are the initial power transmission pumps that pump hydraulic fluid to and from the "box" in the middle in the center on the back in which in the "box" in the middle in the center on the back is an impeller that is to be turned by the hydraulic fluid flow force suction from the first "boxes" behind the driver. The initial power transmission pumps for the super-slipstream bicycle are two quad packs of power transmission units having a gears depth of 13.2 inches.

The paths of least resistance to allow fluid flow to follow around the impeller outer limits (of the second "box") are channels designed in to the impeller on either side of the impeller blade vanes. The force of this hydraulic system is designed on suction and aerodynamics. The hydraulic tubing goes in to and out from the initial pumps (left and right side "boxes" directly behind the driver), on both sides of each pump. The hydraulic fluid lines have bleed valves to purge the air in the lines. A light suction will be applied externally to the suction lines at the bleed valves placed along the suction lines to draw out the air in the suction lines since no reservoir can be placed along the suction lines. The bleed valves will be closed once the air is removed from the suction lines. Hydraulic fluid will be drawn in to the suction lines via the grease seals secondary tubing from the reservoirs through the clearance very slowly at first, a light suction will be applied at the bleed valves to draw out the air in the assembly.

The "cylinders", above and behind the "boxes", have conic sections cut in to them on the inside to funnel the hydraulic fluid in to or out from one line of tubing coming in to the opposite ends of the cylinders from the impeller sides. As the hydraulic tubing reaches the impellers (of which there are two, one is at the rear wheel) the internal diameter of the tubing narrows with respect to the depth of the impeller blade vanes (depth is transverse to the tubing length). This, in effect, creates a velocity change of the hydraulic fluid flow to the impeller blade vanes, which fluid velocity is particular to each impeller. This velocity change is only respective to the upper hydraulic tube of the respective impeller. Two revolutions of the crank with a 24 tooth chainring at the operator develops 16 revolutions at the first impeller and 43 revolutions at the final drive impeller, produced by the hydraulic fluid velocity developed by the first four (quad set of) power transmission units (pumps), then the second set of power transmission pumps drive the rear wheel.

The second pair of power transmission units (left and right rear "boxes") simultaneously deliver hydraulic fluid to and from the second impeller located at the rear wheel. The second pair of power transmission units (secondary power transmission units) have a gears depth of 26 inches.

The [view of the] throttle [is obstructed, it] is [behind a tubing in the view,] next to the rear wheel impeller "box" between the two lines of inflow and outflow tubing. The same principles apply to the hydraulic fluid flow from the second set of power transmission pumps ("boxes") to and from the second impeller only the second impellor is a smaller diameter and the pumps deliver more than 8X the fluid flow volume of the initial pumps with two revolutions of the chainring at the pedals (see formulas) due to the fact that the initial pumps only turn four revolutions per two cranks while the secondary pumps turn 16 revolutions per two cranks of the chainring at the pedals. The result is 43 revolutions per second at the final drive impeller per two crank revolutions of the chainring at the pedals.

A 26 inch diameter wheel turning 43 revolutions per second and a 24 tooth chainring with two revolutions of the chainring in equal time at the crank should require the bicycle to just reach a speed of 200 mph. The wind resistance is estimated to be over 400 pounds, it has not been calculated exactly. Software for Fluid Mechanics Of Inertia is required to calculate the exact air flow force to the vehicle.

Certain measures (tubing wall thickness calculations) have been taken to make sure the tubing does not collapse under the suction load, whether these calculations are correct is as yet untried. As of this moment the only difficulty foreseen is the mangling of the suction tubing under load since the suction tubing will probably all try to become straight as load is applied (this will not happen to the head side of the hydraulic tubing since it suffers no load, only to the suction side). These reservoirs have secondary tubing which feeds hydraulic fluid in to where there is a grease seal as a static seal at each central axis of power transmission unit for each "box" assembly and supplies hydraulic fluid (or anti-freeze and water 50/50 ratio, or sewing machine oil) to prevent the grease seals from sucking air in through the grease seals contact area since there will be a great amount of suction in the gears galleries and impellers housings. The suction force should be minimized because of close tolerances. Hydraulic fluid is supplied to these regions by the reservoirs in both locations which reservoirs' hydraulic fluid is at ambient temperature and pressure and have an opening to prevent draw suction. Once all the air is purged from the assembly the applied force to the hydraulic system should be solid. Bleed valves are placed in strategic locations and a light suction at the bleed valves will disperse trapped air from the main hydraulic tubing. The tubing and power transmission unit casings and plating can be made of polyvinyl-chloride or some other kind of light elastic plastic and their components glued together. Complications of tubing assembly can be remedied with a sleeve and a strategic cut made where the tubing is straight and the sleeve is inserted to cover the cut and the assembly succeeded to by gluing the sleeve to the tubing over the cut. If the entire assembly is made of plastic the tubing can be bonded together with adhesive.

Included in the mechanism are hydraulic brakes with all the required implements. These drawings of hydraulic brakes are not functional as working models in the drawings but are simply representative of hydraulic brakes. Also, there is a twist grip throttle mechanism which throttle is cable and spring operated. The throttle operates by closing off the fluid flow diverted around the final drive impeller so the fluid flow is directed to flow through the impeller gallery: otherwise, with the throttle open the fluid flow is diverted around the final drive impeller and the wheel does not go around. The twist grip mechanism drawing is not functional as a working model in the drawing. These turnkey mechanisms may be purchased separately and are not intended as patentable with the patent application.

The ultimate final drive consists of a connecting chain loop that drives the drive train from the primary impeller input shaft. This assembly has a freewheel at the large chainring which freewheels clockwise (from the left side view) and locks counterclockwise, thus allowing the operator to crank at start-up while the power transmission unit assembly begins functioning. Once the applied apparatuses mathematics take over the ultimate final drive chain loop should apply force to the freewheeling large chainring exceeding the operator's force thereby reducing the operator's applied force to the crank while maintaining and increasing the vehicle's output performance with the throttle while still allowing the operator to crank. This is an unorthodox performance output means but it is mathematically probable and is the basis upon which human powered flight will be developed. This theory is understandable in the mathematics for Fluid Mechanics of Inertia in the power transmission application.

The only three prospects that will prevent this ideal from working up to specifications are suction resistance to fluid flow in the power transmission pump gears, a 90 degree direction change of the fluid flow in the gear pumps, and the fluid flow path in the impeller gallery. The fluid must change direction 90 degrees in the gear pumps twice, although the second time requires no pressure to be applied except to provide its escape. The secondary power transmission must turn 16 revolutions per second at maximum speed, and the primary power transmission turns four revolutions per second at maximum speed. The depth of the secondary power transmission also increases the required fluid flow necessary with respect to the 90 degree change of direction. The primary power transmission is half as deep as the secondary power transmission but there are four times as many of them. Both the primary and secondary power transmissions should apply the same resistance force of suction continuously throughout the instrument's speed range due to the fact that one is twice as deep and turns four times faster and the other is half as deep and turn ¼ as fast but there are four times as many of them.

The fluid flow in the impeller gallery must follow a curved path from the inlet flow area to the draw suction around the impeller. This is constrained by the fluid flow passageways created by the design of channels to either side and above the impeller blade vanes. The remaining clearance of the impeller is limited to only a few thousandths of an inch. The likelihood that the fluid flow will follow the required path around the impellor is probable since it is the path of least resistance.

This instrument is also capable of coasting by which the throttle is released simultaneously (or not) and power is resumed by simultaneously applying cranking and the throttle at speed. The throttle will govern the cadence by regulating the fluid flow volume to pass around the output impeller thus allowing the secondary power transmission to speed up or slow down thereby increasing or decreasing the cadence while maintaining speed.

The operator must consume nutrients in order to apply power.

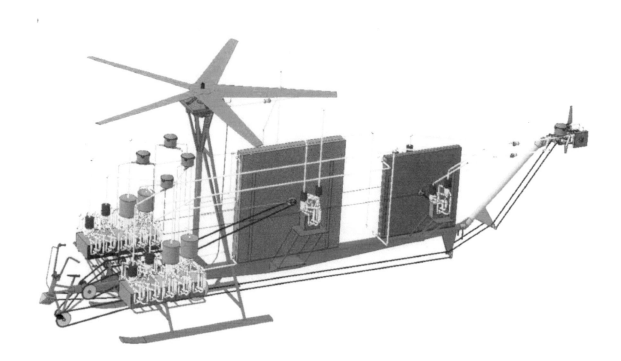

First attempt at human powered helicopter; contains secondary tubing

Unfinished, way too heavy.

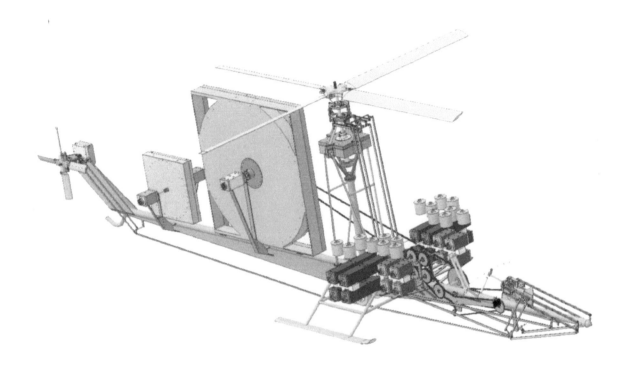

Second attempt at human powered helicopter unfinished andtoo heavy

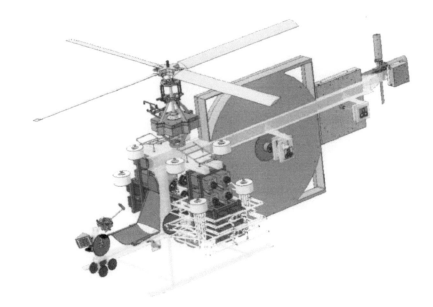

Unfinished, Latest: needs control linkage, chain drive, tubing, throttle…

Chapter 7

Human Powered Helicopter Components and Attributes

The control of the lift blade vanes pitch is now being engineered. This consists of articulated lift blade vanes having ball joints with connecting rods which connect the articulation joining to ball joints mounted in a swash plate which swash plate is articulated around a ball joint in the center mounted to the main shaft to the lift blade. The swash plate is manipulated by control rods with respect to the forward and aft (pitch) blade vanes articulation control by synchronized bell cranks, and the left and right (roll) blade vanes articulation by independent bell cranks. The rotating swash plate is kept from twisting caused by the drag force of the blade vanes during lift, by fingers in the rotating swash plate that extend in to the central ball joint on four sides and by pins in the ball joint that protrude in to the main shaft ring made for them. There is only about 100 pounds of drag force being applied to the rotating swash plate through the connecting rods joining the lift blade vanes to the swash plate. The entire mechanism is operated by control rods and bell cranks to the operator's control stick which is an independent bell crank control lever on two axes with a universal joint at the two levers axes intersection which center will swivel about on a universal joint where the two axes of the levers' rotation intersect. The steering yaw control rotates through the universal joint to the yaw control mechanism to change the tail rotor blade vanes articulation (yaw). The operator moves the handlebars in a circular manner to control the lift blade vanes roll and pitch blade vanes articulation control and the arc levers at the base of the handlebars stem (at right angles to one another) operate the bell cranks that control the swash plate. The operator rotates the handlebars clockwise and counterclockwise to control the tail rotor blade vanes articulation.

The tail rotor blade vanes articulation is controlled in the same way as the previous design of human powered helicopter. The mechanism has not been changed much, only the design appears to be similar. The only drawback remaining in the entire assembly is the flexing of the connecting rods during operation since there now is only a single connecting rod where there were once pairs.

The Power transmission units and the impeller galleries are to be made of polyurethane surrounded by epoxy-carbon fiber with metal inserts for bearing races and containing the gears in the power transmissions. Any other component which may need a metal piece may be applied. Bolts will be inserted with threaded inserts in to the polyurethane and epoxy-carbon fiber for mounting and assembly. The airframe will include bearings to support the impeller gallery output shafts to prevent deformation of the epoxy-carbon fiber and polyurethane constructions, as well the secondary power transmission units will have bearings supported by the airframe. The primary power transmission units' chainrings will have bearings exterior to them to prevent deformation during loading. These bearings are placed at the chainrings where the applied forces are critical. The calculated rate of the impellers for both the primary impeller and the tail rotor primary impeller are four revolutions per crank circuit of revolution. In the previous design of human powered helicopter the rate of impellers revolutions was sixteen revolutions per crank circuit of revolution. Also in the drawings and in the calculations the lift impeller blade vanes profile area is calculated at ¼" square and not ½" square to calculate the secondary power transmission unit meshing gears teeth depth.

The center gears are to be made of titanium, not to say that the center gears need to be tough but that the output shafts need to be strong and are directly tied in to the gears. The center gears and output shaft are made as one piece. The planet gears may be made of aluminum plated with stainless steel. The output shafts for the lift blade and tail rotor are also titanium. The blade vanes hubs for both are titanium. The lift blade spars are titanium. The lift impeller is titanium. The tail rotor impeller can probably be aluminum but if it is titanium it will weigh less although it will cost more. The airframe is 6061-T6 aircraft aluminum but may have to be made of reinforced fiberglass to reduce the weight of the helicopter while hopefully not sacrificing too much strength.

To date, the chain of the chain drive is motorcycle chain, ½ inch width and ¾ inch pitch, which may be stainless steel. The 72 tooth sprocket is a freewheeling sprocket freewheeling clockwise when viewed from the left side. This will allow the operator to crank the pedals crank while the secondary impeller begins to turn. By design calculations the fluid flow to the secondary impeller is greater than is required to turn four revolutions per crank due to the clearance flow volume and the secondary impeller should transfer that extra revolutions force to the 72 tooth sprocket to assist the operator in cranking. However, this may turn in to a runaway mechanism and a throttle will be need to be returned to the assembly. A relief bypass flow may need to be employed to control the runaway problem, if there is one. Sprocket spacing at the primary power transmission units is 10 ½ inches in two dimensions. Sprocket spacing between the 72 tooth sprocket and the primary impeller is approximately 165 inches to allow for chain slack so the chain doesn't break. Sprocket spacing between the 52 tooth chainring at the operator's crank and the flywheel also allows for slack so the chain does not break: however, chain slack in the flywheel drive should be as little as possible to sustain the maximum performance of the flywheel at the operator's crank.

In the drawings there are 36 tooth sprockets at the all the power transmission units. This calculation has been in error and their ratio to the center gear in the power transmission units is 4.5 : 1. With 24 tooth sprockets at the power transmission units the ratio at the center gears will be 3 : 1 which is what the original calculation were supposed to be at. Also the 25 tooth sprocket at the crank will be changed to a 24 tooth sprocket, leaving the 36 tooth sprocket at the crank. All these sprocket changes will affect the spacing between the power transmission units so that the chain will align on the sprockets. I haven't calculated the spacing between the sprockets yet and will not redraw this first drawing of the human powered helicopter but I am going to draw a new version of the human powered helicopter.

Getting the weight down of the helicopter is a primary consideration.

The human powered helicopter is being redesigned. The power transmissions are being located directly under the lift blade; the collectors from the primary power transmissions are being located directly forward of the power transmissions forward of the lift blade axis or rotation. All other components are normal as previously drawn. The aircraft has been shortened by 13 feet seven inches by locating the power transmissions vertically under the lift blade axis. Also, the lift blade primary power transmission units are on the bottom of the stack, the tail rotor primary power transmission units are lighter and are therefore on top of the stack of power transmission units located vertically under the lift blade axis. The primary impellers are both the same from the previous model's design. Their calculations have been adapted. Only the primary power transmission of the tail rotor has been changed with the secondary tail rotor power transmission because the tail rotor is moved from 40 feet out to 27 feet out so the calculations had to be redone and the power transmissions had to be recalculated. The primary impeller for the tail rotor has remained the same from the previous model's design.

Hopefully still, the human powered helicopter will not turn out to be an elaborate ceiling fan. The success of this latest design will be on the lightness of parts. The appearance of the assembly design is representative of the parts' locations. The aircraft must be constructed to meet FAA certification for airworthiness and certificate of airmanship. The end result may not even appear to be the same constructed materials as the drawing portrays them. The airframe may have to be made of plastic because if it is aluminum it will most probably turn out to be too heavy. The output shafts however will have to be Titanium because of their lightness and strength in the places where they are applied, as will the impellers, because of the forces. Torque is very high and forces are great in the places where the Titanium is applied. A minimum of Titanium is ascertained in the design to keep down costs and applied where necessary while aluminum is applied otherwise. The assembly drawings of the human powered helicopter is just a rendering, the actual end result of the parts designs will be different once the values of applied forces in static equilibrium are evaluated for limits of tolerances on parts geometries for shear areas at absolute minimum elastic limit at applied maximum force and the aircraft is constructed to its minimum possible weight. This capacity however may take 100 years to file out of the original prototype (hone to exacting specifications), so I am not expecting miracles out of the original prototype, but it can happen if the right people and today's technology is applied.

This latest version of human powered helicopter has an error in it. The tail rotor impeller is too small resulting in applying too much force to the operator at the crank. As you can see I used previously designed parts to apply to the latest version of the human powered helicopter and the tail has been shortened by several feet increasing the force at the tail rotor without having changed the impeller diameter. Correct this error. The applied force at the operator is about 250 pounds at take off with it as it is.

Hopefully the human powered helicopter once it is finally designed will be light enough to include an electrical system including an alternator, a 12 volt battery, all the accoutrements necessary for driving the 12 volt electrical system and charge the battery, a voltage meter/charging circuit, and amp-meter, an altimeter, two transceiver multi-channel radios with output power/range control, running lights, landing light, strobes, oscillating red light on the tail, an interior light, and an air speed indicator, and circuit breakers. Not much else is really necessary for a human powered helicopter except maybe hydraulic fluid levels in the reservoirs and vacuum sensors in the hydraulic lines, with indicator meters. Warning lights for all the essential points would be considered convenient, and alarms would be convenient as well, as this helicopter is manually controlled there will probably be no need for a computer to monitor anything. Warning indicators should be sufficient.

I haven't finished this last attempt at the human powered helicopter either. It needs control linkage to the lift blades from the operator, the chain drive linkage, the remainder of the tubing which can be drawn using the other human powered helicopters I have drawn, reservoirs, and the self-propeller motor tubing sizes, and it needs handlebars, some other things, and a throttle. This throttle however may be linked between the suction of the primary impeller tubing and its head outlet tubing instead of between the inlet and outlet tubing of the lift blade impeller. Thanks. I will finish drawing this latest attempt to draw the lightest human powered helicopter of the three eventually.

Chapter 8

The Human Powered Helicopter

The evaluation of the power train assembly of the human powered helicopter is the same as for the Extreme Bicycle, only there are a few more evaluations for calculations that must be made for the human powered helicopter. The human powered helicopter has a lift blade and a tail rotor blade which are both powered by impellers. Their calculations are as follows: If the lift blade vane is an airfoil shape and has a straight depth resulting in a rectangular side view profile describing the location of the applied moment of lift is in the diagrams on the following pages.

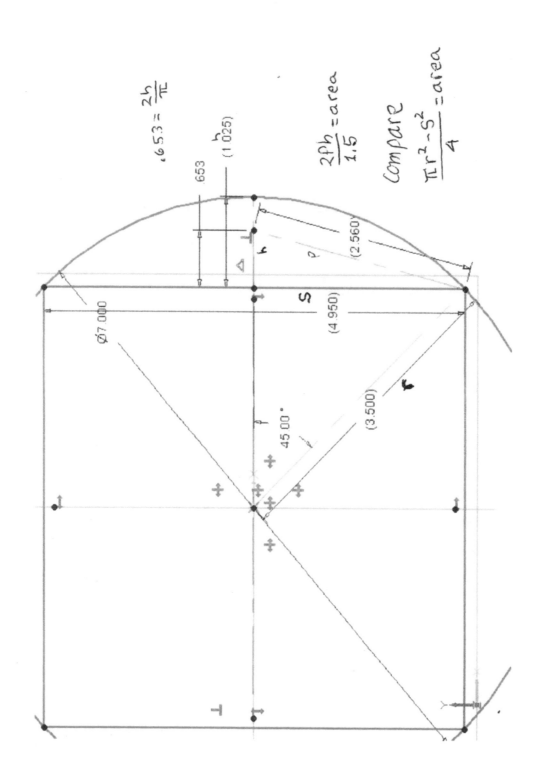

Chord area equation, ½ of a blade vane profile area

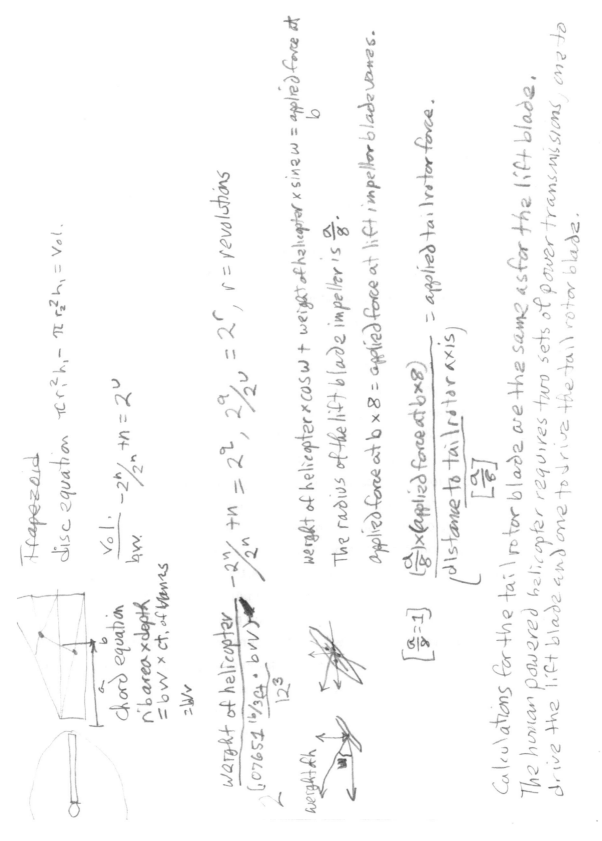

Trapezoid
disc equation $\pi r_1^2 h_1 - \pi r_2^2 h_1 = Vol.$

$\dfrac{Vol.}{b \cdot v} - \dfrac{2^h}{2^n} + n = 2^v$

chord equation
$n \cdot b \, area \times depth = b \cdot v \times ct. \, of \, blades$
$= b \cdot v$

weight of helicopter $- \dfrac{2^n}{2^n} + n = 2^q, \dfrac{2^q}{2^v} = 2^r$, $r = revolutions$

$\dfrac{weight \, of \, helicopter}{(.07652 \, {}^{16}\!/_{3 \, C_1} \cdot b \cdot v \,)} \quad 12^3$

weight $\& h$

weight of helicopter $\times \cos w +$ weight of helicopter $\times \sin 2w = $ applied force at b

The radius of the lift blade impeller is $\dfrac{a}{8}$.

applied force at $b \times 8 = $ applied force at lift impellor blade vanes.

$\left[\dfrac{a}{g} = 1\right] \quad \dfrac{\left(\dfrac{a}{g}\right) \times (applied \, force \, at \, b \times 8)}{(distance \, to \, tail \, rotor \, axis)} = $ applied tail rotor force.

$\left[\dfrac{a}{g}\right]$

Calculations for the tail rotor blade are the same as for the lift blade.
The human powered helicopter requires two sets of power transmissions, one to drive the lift blade and one to drive the tail rotor blade.

Minimal helicopter lift blade calculations… (rough sketch)

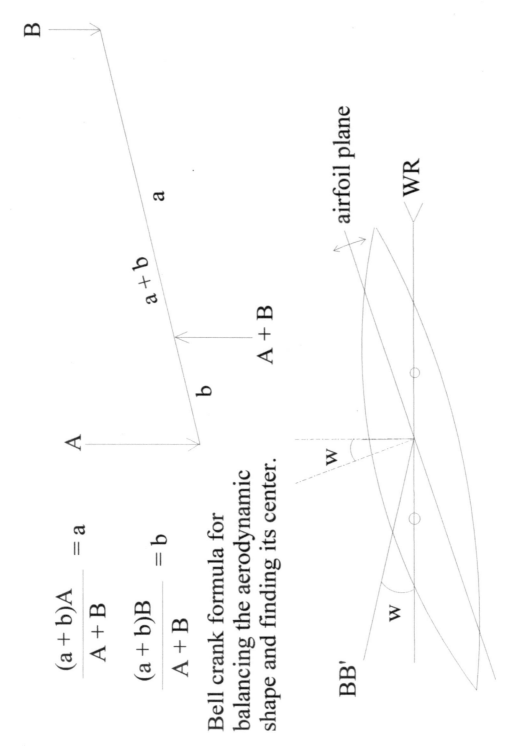

$$\frac{(a+b)A}{A+B} = a$$

$$\frac{(a+b)B}{A+B} = b$$

Bell crank formula for balancing the aerodynamic shape and finding its center.

Areas geometry align their geometric centers at circles on WR. Airfoil plane pivots at lines intersection to align opposite geometric centers.

Bell crank calculations, airfoil vectors in performance

B

a

a + b

A + B

b

A

w

w

BB'

airfoil plane

WR

Once the applied force is found for the lift blade impeller and the tail rotor impeller the same calculations apply to calculate the power transmission gears depths and the primary impellers radii to allow for the sustained operator's input force at the crank. If the force at the crank is out of tolerances simply apply the desired force in ratio with the calculated force in a fraction greater than one (1) and multiply that fraction to the radius of the impeller to equal the new radius of the impeller which will align the limit of tolerances with respect to the operator's force at the crank. The two forces between the lift blade and the tail rotor blade at two crank circuit-of-revolutions per second sustain the maximum operator's force at some desired altitude, say 10,000 feet air density, with respect to the weight of the aircraft (which is required to be minimum, trimmed at every possible place).

The human powered helicopter has Titanium components. The output shafts are Titanium. The secondary power transmissions center gears input shafts are Titanium. The lift blade impeller is Titanium. Components with very high torque force and small radius are Titanium.

The power transmission cases are made of poly-urethane encased in epoxy-carbon fiber with aluminum sleeves for the epicyclical gears and fasteners inserted. The impeller cases are the same materials as the power transmission cases.

The airframe is 6061-T6 aircraft aluminum. The power transmission mounts are welded to the airframe. All the particular components of the airframe are welded to the airframe in their respective places.

Final drafts have yet to be drawn. The specific tolerances of the individual parts have to be finalized before the final drafts can be made.

The parts balancing has to align so the sum of the moments equals zero on the lift blade axis including the operator(s), which operators are all different and each human powered helicopter has to be designed for the variety of operators that there are.

Human powered helicopters will not be inexpensive until they are becoming mass produced and even then their retail price will be significant.

This human powered helicopter is designed to lift a 275+ pound person. The calculations for the lift blade apply 2,431.6 pounds of lift and drag to the lift blade at maximum output. Lift blade revolutions output are shown in the calculations. A calculation of torque applies force to the impeller inside the lift blade impeller gallery as shown in the calculations. The formula for Fluid Mechanics of Inertia applies the required volume of hydraulic fluid to account for the static load of force at the impeller times revolutions plus fluid flow through the relief passageways at static flow volume output at maximum applied force. This sum quantity equals the initial meshing gear teeth volume of the secondary power transmission. The same force at the lift impeller applies at the secondary power transmission. Note that all of the hydraulic forces are in SUCTION. Next, the clearance volume in the secondary power transmission is calculated at the same force, the exponent of the formula for Fluid Mechanics of Inertia is multiplied to the clearance volume and the product is added to the initial meshing gear teeth volume of the secondary power transmission. This is the final volume of the secondary power transmission. Note that all these calculations are made with respect to two crank revolutions at the operator pedals and chainring per second which is equal to maximum output performance, 128 cadence per minute.

The applied force is still the same at the secondary power transmission meshing gears teeth. Now, roughly estimate a circumference for the secondary power transmission impeller and calculate its torque using the force at the meshing gear teeth of the secondary power transmission center gear diameter (preferably the working depth diameter). The force at the impeller radius working depth of the impeller blade vanes is now calculated back through the drive train to the operator. Once the desired Operator force is applied divide a percentage of the desired force at the operator (say, 67%) into the resulting force of the calculations at the operator and multiply the quotient to the diameter of the impeller and the product should result in the desired diameter of the impeller resulting in the 67% required operator's force through the drive train at the pedals center axis. The 67% applied force will equal 100 minus the tail rotor force divided by the lift blade force times 100 equals the desired percentage of primary power transmission operator's force.

The applied force at the secondary power transmission is now calculated in torque to be the desired force at the resulting diameter of the secondary power transmission impeller diameter. Use the formula for Fluid Mechanics of Inertia to calculate the fluid volumes of static load volume of the impeller blade vanes times revolutions plus the relief passageways volume at static load exponent, and this will be equal to the primary power transmission initial meshing gear teeth volume. The applied force at the secondary power transmission impeller conveys to the primary power transmission and is the same force. Now calculate the clearance volume in the primary power transmission and using the force at the secondary power transmission impeller blade vanes and the formula for Fluid Mechanics of Inertia calculate the static volume of the clearance volume of the primary power transmission at the applied force using the exponent of the formula for Fluid Mechanics of Inertia and add the product to the initial meshing gear teeth volume of the primary power transmission initial meshing gear teeth volume. This will be the final meshing gear teeth volume of the primary power transmission.

The Tail Rotor

The applied suction at the lift impeller will make the helicopter want to rotate the airframe clockwise looking down from the top. Again, torque applies at the radius of the lift impeller with respect to the axis of the tail rotor. Keep in mind that all forces are applied with zero motion although motion is accounted for as revolutions in calculations. The same formulas apply with respect to the tail rotor blade vanes and with the lift rotor blade vanes although the tail rotor blade vanes will pitch to make changes for yaw control, increasing and decreasing or reversing the amount of lift plus drag force created by the tail rotor blade vanes.   The entire force at the tail rotor (lift + drag) is used in calculations to determine the dimensions of the impeller and the power transmission meshing gear teeth volume of the tail rotor power transmission assembly exclusively.

The impeller for the tail rotor is calculated to turn the required output revolutions by minimizing the frontal area profile of the tail rotor impeller blade vanes while managing the secondary power transmission of the tail rotor impeller so the impeller of the tail rotor secondary power transmission and primary tail rotor power transmission are all kept limited to minimum tolerances. Varying the diagonal frontal area profile of the impeller blade vanes and the impeller's diameter will align the tail rotor secondary power transmission to minimum tolerances. The applied force at the tail rotor impeller (lift plus drag (lateral and transverse)) is calculated back through the drive train to the operator and added to the initial operator's force (for example: 33% tail rotor force + 67% lift blade force = operator's force). The sum of the two forces at the operator should be the desired operator's force. The diameters of the secondary power transmission impeller and tail rotor secondary power transmission impeller must be aligned to achieve the desired operator's force.

The sum of the two forces is applied at the operator pedals and must be within the operator's tolerances. Hydraulic fluid should be sewing machine oil or some likeness of thin lubricating oil of low viscosity, perhaps linseed oil but must not break down over time and not be hygroscopic. Most parts are to be made of elastic plastic or some other durable light material that is inflexible. The control linkage is a simple gear driven bell crank and connecting rod assembly.

The tail rotor spins counterclockwise looking at the tail rotor from the left side of the helicopter. The tail rotor mechanism consists of a ball jointed bell crank which turns a screw that operates a lever with a forked end that moves a slider (the internal component of which spins on the tail rotor output axle) that operates the blade vanes to pitch, whereby the lift and drag of the tail rotor are controlled and so yaw occurs. The tail rotor blades turn on an axle that is engaged with the tail rotor, and output impeller. The screw threads of the screw that operates the lever are left hand threads for the required blade pitch with the steering and control linkage.

The Secondary Hydraulic System

All of the hydraulics have reservoirs. The secondary hydraulic system operates by syphoning hydraulic fluid from the reservoirs. The purpose of the secondary hydraulic system is to prevent air from leaking in to the entire assembly at the axes of rotation through the grease seals. Since there is suction throughout the entire assembly during operation under load there is continuous suction at the axes of rotation at the grease seals and hydraulic fluid will be sucked in from the reservoirs through the secondary hydraulic tubing instead of there being air sucked in at the grease seals. By applying secondary hydraulic tubing to suck hydraulic fluid under suction force the possibility of sucking air in at the grease seals is eliminated. Hydraulic fluid is continuously replenished in the reservoirs by the continuous recirculation of hydraulic fluid in the assembly.

There are six reservoirs on this helicopter.

Pitch and roll are controlled by the operator leaning forward and backward and from side to side. There are four locations for counterweights at the rear of the helicopter. The counterweights are bullet shaped with a threaded end and hang down by their threaded end. The counterweights will probably be to offset the weight of the hydraulic fluid. A counterweight to offset the weight of the pilot has not been added at this time. No account has been taken for lift blade vanes pitch control at this time.

There is a throttle. The throttle is hand operated by a twist grip mechanism at the right hand on the handlebars. The throttle controls the flow of hydraulic fluid to the lift impeller by closing off the bypassing fluid flow bypassing the lift impeller gallery by channeling it through the throttle and allowing more hydraulic fluid to flow to the lift blade impeller gallery thereby applying more hydraulic fluid flow to the lift blade impeller gallery increasing the force of flow to the lift blade. The operator may crank the pedals and vary the cadence while operating the throttle and control the lift blade output revolutions and the cadence by manually operating the throttle, to be able to ascend and descend in a more controlled manner and at a more comfortable cadence.

The control of the lift blade vanes pitch is now being engineered. This consists of articulated lift blade vanes having ball joints with connecting rods which control the pitch joining to ball joints mounted in a swash plate which swash plate is articulated around a ball joint mounted to the main shaft to the lift blade. The swash plate is manipulated by control rods with respect to the forward and aft (pitch) blade vanes pitch control by a bell crank, and the left and right (roll) blade vanes pitch by independent bell cranks. The rotating swash plate is kept from twisting caused by the drag force of the blade vanes during lift, by fingers in the rotating swash plate that extend in to the central ball joint on four sides and by pins in the ball joint that protrude in to the main shaft ring made for them. There is only about 100 pounds of drag force being applied to the rotating swash plate through the connecting rods joining the lift blade vanes to the swash plate. The entire mechanism is operated by control rods and bell cranks to the operator's control stick which is an independent bell crank control lever in two axes with a head tube at the two levers axis which axis will swivel about on a universal joint where the two axes of the levers' rotation intersect. The steering yaw control rotates through the universal joint to the yaw control mechanism to change the tail rotor blade vanes pitch (yaw). The operator moves the handlebars in a circular manner to control the lift blade vanes roll and pitch pitch control and the arc levers at the base of the handlebars stem (at right angles to one another) operate the bell cranks that control the swash plate.

This concludes the explanation of the human powered helicopter.

Chapter 9

The Propensity Of The Self Propelled Motor

As the rocket motor seemed to be self-propelled so the self-propelled motor would seem to propel itself as well. Thus, the mechanics for self-propulsion of motion in a motor is presented in a mechanism in these drawings; these drawings are placed in order of evolution:

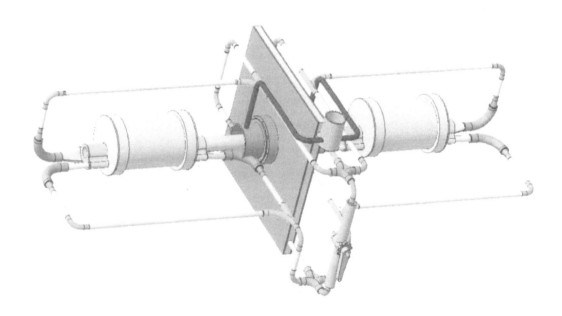

Evolution of the self-propelled motor, 1st drawing (CAD software)

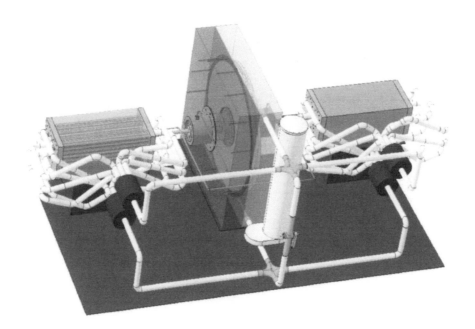

Second drawing (CAD software)

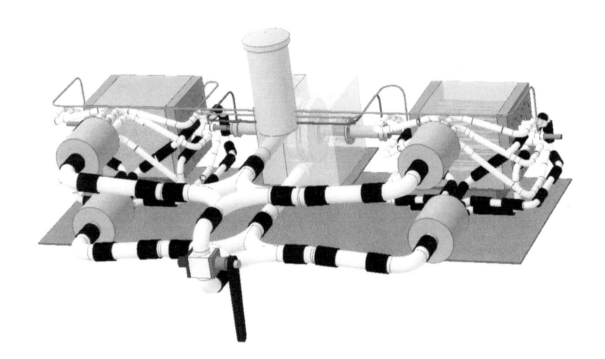

Final draft (CAD software)

The mathematics for the self-propelled motor apply with respect to the effect that multiple revolutions are applied in product (multiplication) with respect to the coefficient of the impeller gallery volume and the number of revolutions desired exceeding one (1) revolution of the impeller. By calculation the meshing gears area is divided into the impeller gallery volume less the impeller x the number of revolutions = the meshing gears depth. The meshing gears radius is less than the impeller blades vanes radius by a significant ratio. Preferably the meshing gears depth makes the meshing gears a square by side view profile with respect to the radius of the meshing gears compared to the impeller blade vanes radius and applied calculations for the depth of the meshing gears. But the square doesn't really make all that much difference, the object is to get the hydraulic fluid to suction freely to the power transmission throughout the assembly, from the head flow of the meshing gears to the reservoir, from the reservoir around the impeller blade vanes in the impeller gallery and back to the power transmission suction inflow meshing gears and around the gears to the head flow meshing gears in a continuous cycle, the excess head hydraulic fluid being bypassed by the throttle valve before it gets to the reservoir back to the suction lines allowing for acceleration and deceleration of the motor in performance.

Since the moment is in the flow force from the suction at the meshing gears inflow and the flow of the hydraulic fluid around the impeller blade vanes is applied in multiple revolutions with respect to the revolutions of the impeller's actual turning rate, the flow force on the impeller blade vanes applies torque to the meshing gears which applies suction force to the hydraulic fluid in the meshing gears which force is greater than the force at the impeller blade vanes because it is applied in torque and the meshing gears radius is smaller than the impeller blade vanes radius, the throttle valve will bypass the excess flow of meshing gears suction of hydraulic fluid from the head flow back to the suction flow standing idle sustained while closing the throttle will reduce the head flow to the suction meshing gear teeth and increase the suction through the impeller gallery and the head flow to the reservoir so more flow goes to the impeller in the impeller gallery applying more force to the impeller blade vanes increasing the torque and therefore the force at the meshing gears teeth and as a result creating acceleration in the internal working parts as well the reverse is true creating deceleration in the working parts by opening the throttle valve allowing the head flow to pass to the suction meshing gears teeth suction lines and not to the reservoir. The force applied at the suction meshing gears teeth is the same force that is applied at the impeller gallery port area through the hydraulic line, so there should be torque.

With all due respect this may be the only self-propelled engine that may ever work, if it works. The reason the author has concocted this concept is that it *contains* all the hydraulic fluid and does not spill any of it out like the designers of self-propelled engines of old spilled all their hydraulic fluid out in their drawings of self-propelled engines.

As you can see in the evolution of these designs the diameter of the tubing aligns, from straight tubing of one single diameter to tubing of the sum of the areas of the tubing in line with the exception of the area of the ports at the impeller gallery. The alignment of the areas allows the hydraulic fluid to flow without adding any additional force of acceleration from having to channel a sum of areas in to an area equal to 1 area/sum of the areas x (sum of the areas) = 1 area, which would cause the hydraulic fluid from the sum of the areas to have to accelerate in to the 1 area creating extra force to have to be applied at the suction meshing gears teeth reducing the efficiency of the performance of the self-propelled motor.

Now the tubing size is equal to the sum of the areas. The hydraulic fluid flow is constant. There is no acceleration force to add. The head force no longer has an added force from having to accelerate the hydraulic fluid. Deceleration of the hydraulic fluid would not have applied any useful effective force in the self-propelled motor because the force of deceleration would only have applied a load to inflate the draw force meshing gears with hydraulic fluid as the gears turned. Still, eliminating the acceleration of the hydraulic fluid eliminates the restriction of the forces of acceleration from a single diameter of tubing.

Suction is applied throughout the hydraulic fluid on the suction lines side of the system to the inlet port area at the reservoir just before the impeller gallery no matter what the diameter of the tubing is, the suction is still the same at the inlet port area no matter what the diameter is between the inlet port area and the meshing gears suctioning the hydraulic fluid. If there is a gas in the lines there will be spring/bounce in the suction force. The gas must be removed for the force to be solid.

The head side of the meshing gears only pushes the hydraulic fluid to the reservoir it does not have any pressure on it. Only the viscosity of the hydraulic fluid will apply reaction force in the meshing gears teeth as the hydraulic fluid is squeezed continually between the teeth of the meshing gears to be pushed to the reservoir. This squeezing effect is continual in the performance of the self-propelled motor of this kind, as is the suction flow of the hydraulic fluid continual simultaneously.

The dimensions of the impeller blade vanes can be changed. Its radius and depth can be varied to accommodate a limit with respect to a certain size of inlet and outlet port area. This size change however will affect the meshing gears depth. All the respective radii are taken into consideration. So, if you are going to make the impeller blade vanes big and fat you will probably have to change the radius of the impeller so the meshing gears depth will make the side view profile of the center gear a square when its depth is calculated. If the impeller blade vanes are big and fat the radius of the impeller will be really small if the side view profile of the meshing gears is still a square. Surely there is a point of diminishing returns on the radii which can be applied when designing this self-propelled motor. I will leave getting the most out of this motor for the future of mankind and their ability to develop machines to their state-of-the-art capability.

To calculate the size of the power transmission meshing gear teeth volume the size of the impeller has to be described by choice. The application of fluid flow volume to the impeller can either be by a multiplier, by a multiple of two, or by a force in pounds, preferably in a multiple of two simplifying the analytical procedure. In this case 24 is the multiplier used to calculate the gears depth. Without using the formula for Fluid Mechanics of Inertia and the density of the hydraulic fluid, calculate the volume of the impeller gallery less the impeller volume, impeller bracket gasket and impeller bracket and multiply by

24. Calculate the impeller blade vanes volume and multiply by 24 x 1.5. Divide the sum of the total volumes by the area of the meshing gear teeth of the power transmission, using one area and multiply that one area by the number of meshing gears. This is the final meshing gear teeth volume for a 24 multiplier power transmission. Then divide by two power transmissions. Fabricate gears (center and planet gears are equal depth) according to these calculations. Fabricate the power transmission casing to the same depth. A +0.007" gasket separates the power transmission casing from the power transmission plate. The power transmission plate contains the ball bearings for the center gear and the pilot bearings for the planet gears and the openings for the fluid flow tubing, and bolt holes for the power transmission grease seal. Assemble the power transmission. The grease seals fit over the center gear output axle and bolt to the power transmission plates with a gasket between them. Assemble the power transmission to the mount.

Assembling the impeller assembly

In this case apply a 3 inch radius to the impeller. This is the working depth of the impeller blade vanes. These impeller blade vanes are ¼ inch by ¼ inch in the profile view. The total radius of the impeller is 3 1/2 inches at the impeller blade vanes. The impeller is ½ inch thick at the impeller blade vanes but the blade vanes are only ¼ inch thick leaving 1/8 inch to either side of the impeller blade vanes for clearance volume. Apply a 3 1/2 + 1/16 inches radius to the outside of the impeller and extrude to depth to one side approximately 1 inch deep, this leaves 1/16 inch clearance above the impeller blade vanes. To the other side apply the same radius to an extruded depth of ¾ inch. The depth of the deeper side of the impeller will have to be adjusted as the impeller casing is fabricated. Fabricate the casing of the impeller to include 0.007" radius clearance (+0.014" diameter) for the impeller to fit into the casing. The grease seal and gasket fit over the center gear input axle, then the impeller casing fits over the input axle of the center gear and the bearing goes in and bracket fits over the end of the center gear output axle and is attached with an E-clip. The impeller brackets attach to the center of the impeller on one or the other sides with a +0.007" rigid gasket between the bracket and the impeller to keep hydraulic fluid from leaking out through the bracket at the center gear input axle. The bearing rides on the bracket. Clearance is machined in to the impeller casing for the bracket, gasket and bolts, and a space is machined for a bearing in to the impeller casing. The remaining bore through is +0.014 " the diameter of the bracket diameter that will stick through the impeller casing. The depth of the casing for the impeller will allow the fluid flow ports to be drilled out to allow for heavy diameter and heavy wall thickness tubing, this is in the event that if there is a given load to the self-propelled motor and the suction force increases the tubing will not crack. The holes drilled in the casing must have enough material to one side (the open side for the impeller to enter) so the drilling operation will not deform the material at its thinnest area to the open side. Therefore, the impeller depth to one side must be aligned to the plane of the impeller casing depth while the impeller blade vanes and clearance area and fluid flow ports centerlines must align keeping the +0.007 inch clearance all around.

The impeller grease seal and gasket go on the other center gear input axle, then the impeller plate fits over the same center gear input axle and the bearing goes in, and the other impeller bracket goes on the end of the center gear input axle and attaches with an E-clip. A 0.007" rigid gasket goes between the bracket and the impeller and the brackets are then bolted to the impeller. A +0.007" gasket goes between the impeller casing and the impeller plate and the two halves of the impeller assembly are fitted together. The same bores are taken for the bracket to the impeller on the plate side including the bearing. The entire assembly is mounted on the mount as construction is proceeding.

Assembling the tubing and self-propulsion

The elbow tubing is placed first at its respective angles according to the drawings. The straight tubing and elbow tubing are joined by rubber sleeves with hose clamps and the tubing are pressed as close together as possible to prevent an aneurism from occurring in the rubber sleeve at the gap in the tubing where the tubing comes together. The tubing is butted together. The tubing fits in to the collectors with rubber sleeves. The tubing is glued or soldered in to the impeller plate fluid flow ports. The throttle also is fitted with rubber sleeves. The secondary hydraulic tubing is fitted from the grease seals to the reservoir and fixed in place. Hydraulic fluid is drawn in to the grease seals from the reservoir through the secondary tubing as suction is drawn by operation of the impeller and with load as operation applies suction. Hydraulic fluid is recirculated back in to the reservoir. The throttle tubing is assembled the same way. The secondary hydraulic system does reduce the probability of power slightly but it is the only way to seal in the hydraulic fluid. The reservoir is filled, a crank is inserted in to the outboard end of the center gear (on the right side) and the crank is turned clockwise as the throttle is closed, hydraulic fluid flow is channeled in to the impeller gallery by suction and away from the throttle (throttle closing), the amount of force of suction being applied at the meshing gear teeth is the same at the impeller suction fluid flow port; there is a 3:.95 inch radius ratio of torque where the radii of the two forces are equal (between the impeller blade vanes working depth and the meshing gear teeth circumference). As the throttle is closed the flow volume to the impeller increases increasing the torque while the forces remain equal. The force of torque being applied would seem that the sum of the moments greater than zero is achieved. This device can be varied by applying the gears depth with respect to the radius of the impeller (changing the multiplier), the impeller clearance volume around the impeller blade vanes, and the blade vanes volume. Some other variables may apply like gear teeth count for smoothness and minimizing friction, but the author is unable to constitute a reasonable faction for why this self-propelled motor will not work except that self-propelled motors are notorious for not working.

Update

Lately the tubing diameter has changed. The areas of the tubing have been equalized so the sum of the areas of the six inlet ports of the collector are equal to the area of the outlet port of the collector. The related tubing area have all been made the same diameter. The only change in area occurs at the port area of the impeller gallery at which point the fluid flow is required to speed up anyway. This tubing size change relaxes the hydraulic load on the fluid flow. Still, hydraulic fluid flows to the inlet and outlet ports of the impeller gallery only with less haste and there is virtually no pressure or suction load compared to what is inside the impeller gallery. Pascal's Law of areas in hydraulics makes an equal amount of volume displace from six in to one out with equal area for both. I suppose the displacement distance ratio is 1:1 if the areas are equal.

Here is a problem: Suppose head and draw forces are equal at the meshing gear teeth at 7 + 7 = 14 at the radius of .95. The radius of the impeller is 3 and has a force of only the one draw force of 7. 3 x 7 / .95 > 14 x .95 / 3. Even if there is no head force to the impeller there is rotation in the torque lever arm. All the head force is expelled in the reservoir. Even more probable is there is very little head force at the meshing gears teeth head outflow. Only the viscosity of the hydraulic fluid will introduce head force. Figure out some way to prove that this motor applies with respect to the sum of the moments equal to zero and I will admit that it doesn't work. No friction is included in this equation but will be included in the actual working model.

For reference, all the previous "perpetual motion" motors I have seen in my life growing up spilled all their hydraulic fluid out in their attempt to perform; my self-propelled motor's hydraulic fluid is all contained and none of it spills out. Also, this is not physics this is statics.

It seems impossible but even when the moving parts are held in static equilibrium and are kept from moving and the fluid forces are applied at their respective areas the torque equation still applies the sum of the moments greater than zero with the throttle fully closed. It may be required that the throttle be nearly fully closed for the machine to operate. With the moving parts released from static equilibrium and allowed to move freely and the throttle is closing mostly all of the way, and the parts are applying their respective forces all simultaneously, what could be causing the operation to fail? Discover the solution to this dilemma and resolve it by conforming the device to accommodate a solution to its operating principles. There must be some cause for why it won't work, the cause just hasn't been discovered yet. The solution for why it won't work has not been found. Only the areas ratio of the impeller gallery inlet and outlet ports and the meshing gear teeth inflow and outflow ports is left to question.

Even on the atomic level an atom or molecule of hydraulic fluid applying a force in the inflow and outflow of the meshing gears teeth (twelve atoms or molecules respectively (twelve inflow and outflow ports individually at the meshing gears teeth)) will apply six molecules or atoms at the inlet and outlet ports of the impeller gallery at the same force that is being applied at the meshing gears teeth and the sum of the moments will be greater than zero. Remember that the equation being applied to the impeller blade vanes is the equation for fluid mechanics of inertia: hydraulic fluid density x blade vanes displacement volume x 2 raised to the power of the quantity of the applied force divided by the quantity of the hydraulic fluid density x the blade vanes volume close quantity, minus 2 to the nth power, then divide by 2 to the same nth power, then close the exponent quantity and subtract the exponent quantity by the same n value, equals the hydraulic fluid density x the blade vanes volume x 2 raised to the power of p, which product is equal to the applied force in pounds. The applied force is the same force that acts in the meshing gears teeth, and if it is double for inflow and outflow for it to be the same force at the impeller gallery inlet and outlet ports simultaneously the equation for the sum of the moments is still greater than zero. The calculation for the meshing gear teeth volume is taken from the calculations from the impeller gallery volume to the meshing gears teeth volume as written at the beginning. This is complicated and the subject is beginning to go around in circles, no pun intended. Good luck. It is proving very difficult to make this self-propelled motor not work.

And, once again, the author is financially incapacitated and cannot function financially as an inventor: therefore, the self-propelled motor is being submitted for copyright. Thank you for your interest in the author's ideas and discoveries. Perhaps one day these ideas and discoveries will serve their useful purposes. No financial progress can be made. The author is sickened by the prospect of not being able to make any financial progress.

Even if the ratio of the sum of the impeller gallery inlet ports areas divided by the sum of the large areas of the four collectors is multiplied to any force and then multiplied to the ratio of the two radii in reciprocal and then one of the two is subtracted from the other while the fulcrum is the axis of rotation of the center gear, still the sum of the moments do not equal zero. Forcing the sum of the moments to equal zero is proving difficult.

Pascal's Laws show that a large area will sweep a short distance a particular volume and the same swept volume will sweep a long distance through a small area. The radii of the two areas in this case are in consideration: the port areas of the meshing gears teeth and the impeller blade vanes. Is the applied force to the impeller blade vanes diminished because the impeller gallery inlet and outlet port areas are smaller than the sum of the meshing gear teeth areas? Where did the applied force go? Was it dissipated somewhere? If so then where is it? If not then the applied force applies at the impeller blade vanes volume, area doesn't matter. Apply the formula for fluid mechanics of inertia, force is applied at the meshing gear teeth to the hydraulic fluid in static elastic equilibrium in zero time present simultaneously with respect to the simultaneous applied force to the impeller blade vanes volume on both radii. Notice how many variables can be changed: 24 can be changed to 16 or 8 or 12, these variables have been mentioned already, and the whole mechanism can be redesigned to accommodate the change in variables. Close the throttle and crank the crank, the self-propelled motor should start, if it doesn't I don't know why not. Keep trying.

Chapter 10

Gear Design Strategy

A necessary component in the construction of the human powered transportation means is the sprocket. The following drawings give a quick overview of the design parameters for drawing sprockets on CAD software. It's easy so don't be too concerned. Gears need to be designed for inside the power transmissions, and sprockets need to be designed for the chain drive:

If the distance along the circumference between two gear teeth is equal to an American Standard units constant increment of length, then the formula pi times the diameter divided by the number or teeth equals that increment of length, whether it be 1, or 1/2 or .5, or 5/16 etc. With algebra the formula can be manipulated with the diameter as the unknown and applying the number of teeth will vary the diameter. The American Standard increment is known as the circular pitch. Other formulas for spur gear teeth can be found in the Machinists Handbook.

You want this distance to equal .5 (see example)

Varying the teeth count will vary the degrees and the distance with respect to American Standard Units.

$$\frac{pi \times D}{T} = \text{whole number}$$

example: $\dfrac{.5 \times 12}{pi} = 1.90986\text{" diameter}$ $\dfrac{T}{D}$

You have the whole number of 12 gear teeth, T.

As it turns out, the outside diameter of the pinion gear is 2.2281 inches and the outside diameter of the driven gear is 2.54648 inches because the driven gear has 14 teeth. The driven gear diameter is the same as the pinion gear outside diameter, 2.2281 inches, so the centers distance is one half the diameters of either gear, or 1/2 x 2.54648 + 1/2 x 1.90986 = 2.06898 inches.

Varying the gear teeth count will not change the whole number. D is the unknown.

.5

30.00°
Circular
pitch

D

Gears have three diameters. There is a diameter between where the gear teeth bottom out, and there is a diameter at the top of the gear teeth, and there is a diameter in between those two diameters. The mid diameter is where most of the initial calculating begins. It is also the diameter to determine the centers distance from. Finding the middle circumference's diameter is where I have had the most trouble starting my gears drawings.

The Machinists Handbook has the gear cutter numbers for the gear teeth count that you will need to use. The total gear depth is how deep you should machine the gear teeth.

Gears design parameters

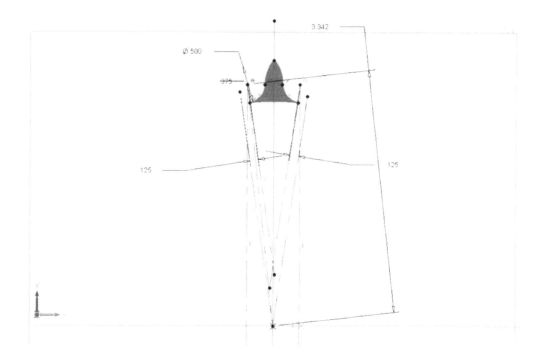

Sprocket Tooth Design Parameters For CAD, 28 tooth sprocket

Some techniques exist in the designing of a sprocket tooth in CAD but it is fairly simple. It takes practice but is not difficult.

Chapter 11

The Exploits Of Human Powered Flight

The problem with having a gasoline engine, while flying a helicopter is running out of gas and not having a place to land. A human powered helicopter is not like that, the problem with the HPH is you will run out of strength and all you will have to do is set down and wait and your energy will charge back up all by itself. You may also need to eat something but mostly just resting for a little while is enough to catch your breath and get your energy back to further your journey.

The prospects of going after the indigenous psychoactive intoxicant wildlife (i.e. the marijuana plant, the psilocybin mushroom, the peyote cactus (mescaline), the coca tree (cocaine), and whatever else I can't think of) is inevitable. People on human powered helicopters will scour the countryside for these plants and the Earth would be stripped bare of them were there no law enforcement case providing officers with human powered helicopters to prevent such a catastrophe, when actually the law officers will probably use the human powered helicopter to go in search of these indigenous life forms to destroy them themselves. Once the human powered helicopter comes of age mankind will seek out and lay waste to the wilderness of its indigenous psychoactive wildlife. However, this is not a bad thing. Once it is ascertained that the indigenous intoxicant wildlife is in danger of becoming extinct the Government should affect the endangered species act or whatever is necessary to protect the indigenous psycho-active intoxicant wildlife, which is counterintuitive according to the present state of intentions of the present Government. Otherwise they will just wait and let man extinguish the plants from off the face of the Earth. Mankind is not smart enough to think before he acts when he craves these substances, dead or alive, especially in the United States where there is much abundance.

Human powered flight will make it possible to raid farms for food, although it would seem silly to form a raiding party with human powered helicopters since a human powered helicopter will be quite expensive and it doesn't seem likely that anyone who can afford one will be suffering from any kind of starvation prospects and would have no need to form a raiding party: still, in the future raiding parties may occur from natural disaster or some act of God. Farmers will need huge field sized nets to cover their crops so that individuals on human powered helicopters coming to raid their farm cannot swoop down and gather from the farmer's fields. Farmers shall not be allowed to shoot people however, they will only be within their means to call the police. The mesh of the nets will have to be fine enough to keep human predators out of the harvest, cutting through the nets covering a farmer's harvest could probably be a more serious offense, and getting caught one could do time, pay a fine, or both, for cutting a farmer's harvest cover net.

Telephone lines will also be a bother. What will become of the telephone lines and power lines and electrical wires running between the telephone poles in and between cities and towns? It would seem feasible to string a fiber optic line from telephone pole to telephone pole that light up in the evening and stays lit all night so that the locations of the telephone poles and wires are obvious to pilots of human powered helicopters. The event would have to be funded by indiegogo or kickstarter or something like that. The military bases will have to do similarly and string a fiber optic light line around the military base to keep out the non-military public, or think of some other kind of means to deter people from flying in to a military establishment unawares.

The first human powered helicopter will have to obtain permission from Congress to fly over, into, through, upon, and under Federal Land Open To The Public to be able to leave a private restricted area in the first place.

Homes will have their garages built on their roof to accommodate the availability of human powered flight in the future. City skyscraper structures may have landing platforms built outside the offices of individuals for the purpose of convenience of access to one's human powered helicopter and/or office.

Flying this human powered helicopter will be elementary. Considering the modern day traffic or automobiles on interstate highways and during the rush hour, they are all bunched together on a single strip of pavement. Human powered helicopters will free man from the pressures of city life and grounded traffic opening up the sky with all of its innumerable altitudes and directions which can be taken on a whim. Establishing freedom in the sky for mankind in this country will require the Government yielding to the prospect that in time scenic routes will precipitate out and can be adopted as major throughways. For there to be determination to control the trajectories of our flights by imposing on man to dress and cover in flight, or to fly a designated flight path laid down before a scenic route has precipitated out, or to cause single file flying, bottlenecking et cetera, will need to be outlawed or prohibited in some way thereby eliminating traffic jams and a major cause of accidents in the sky. The deregulation of human powered flight should be a Right (as is prohibition) established by Government to free the people from catastrophe and so that our wings can never be taken away from us by any justification of justice. Owning a human powered helicopter should be a Right under the Fourth Amendment, as a keeping and bearing of arms and a rule needs to be applied that prevents the police from the confiscation of a human powered helicopter as a weapon is confiscated during an arrest.

As the nobleman goes along on his way, so the nobleman shall be free on his journey according to the Scriptures of the Holy Bible. Traveling in to wildernesses on a human powered helicopter should not be an excuse for law enforcement to pursue after the individual with the intent of apprehending them on any grounds without any probable cause, preponderance of evidence, burden of proof, or eye witness of a crime having been committed by the operator of the human powered helicopter only by the suspicion that an operator of a human powered aircraft is a drug addict. Appearing to not be doing anything does not justify apprehending someone in a National Park and imposing on the person to have to serve time in a mental hospital because he looked like he wasn't doing anything, in the National Park, should be investigated and set right. Since when does appearing to be doing nothing justify imprisonment in a mental hospital? Occupying one's free time in a National Park by one's self is not a crime, nor is it blatantly insane, or any intent to commit a crime and does not justify so much as even an approach by a law enforcement officer or Park Ranger. Flying to the mesa of a chimney in a human powered helicopter out West would deter any such approach, and sit there and appear to be doing nothing for the rest of the afternoon, you probably won't get approached by a cop on the ground up there and being taken in to custody and carried off from the top of a mesa tower, to serve time in a mental hospital, may prove difficult.

The objective by Government securing the sky as a National Preserve renders the requirement of a radio transceiver with multiple transmitter power settings which can be used to communicate short distances as well as long distances by the flip of a switch without shouting across the sky to another operator or shouting at all, because it is silent up in the sky except for the wind blowing and rain and thunder and the only shouting to date is the birds shouting to one another, birds are always shouting anyway, and the preservation of the silence should be kept secure, the sky should be kept as wild for future generations as it is now. Painting the clouds with food coloring should also be restricted to special occasions and the clean water act should take a precedent in case of overenthusiastic cloud painters making the water taste bad eventually, and change color from clear to a mucky brown from the mixture of all the different colors of food coloring.

It will be impossible for the operator of one human powered helicopter to reach out and touch the operator of another human powered helicopter, as you can determine by the allowances of the blades locations during flight when approaching another human powered helicopter.

Still, it is not a wise decision to sow seeds while aloft in a human powered helicopter unless farming on one's own land, with the exception of sowing seed balls. Leave sowing wild seeds to the birds. Birds love to eat seeds and expel seeds, and there would be plenty of seeds to be sown if mankind had not got it in his head to lay waste to the indigenous foliage life forms over all the Earth in the meantime, on his human powered helicopter. Sit there on your sack of seeds and just smile and wave. Then there is the issue of where to land: in National Parks manmade "trees" can be built in remote places that will support human powered helicopters as landing pads and instead of landing on the earth and causing any erosion they can land on the "trees" provided for them: otherwise, it probably would be best to land on rock or concrete to prevent soil erosion in the wilderness, despite the fact that the human powered helicopters will probably only be able to land on flat surfaces including the water if so equipped. These "trees" will be made of materials and construction designs to support human powered helicopters like landing on fixed and rigid leaves. With pontoons you will be able to go fishing, and fly home to your lovely family and fix what you caught to eat.

Getting chased by the cops will change too. From once long ago there was running and trying to get away being pinned to the Earth by gravity. Now there will truly be flight, and fight or flight will change to fight and flight, and eluding pursuit will be something to deal with. Dog fights may erupt on rare occasion between the assailants and the police.

The Holy Bible makes a reference to a holy multitude coming so dense that the light of the sun was darkened and the sound of their wings was as the voice of God. Should it be possible for everyone in this country to own a human powered helicopter then a mighty fighting force will we be in the event of belligerence, should the military need assistance. As well, human powered flight will make mankind capable of migration in the event of a solar system-wide cataclysm.

There should also be outlawed shooting wildlife from a human powered helicopter, including coyotes. The author does not think anyone should shoot any animal from an aircraft, how unfair is that. The author believes hunting from an aircraft is already outlawed in the United States but I don't know for sure. The author has seen a video of shooting coyotes from ultra-lights or something like that years ago on TV. The fact is the author cannot even fly the human powered helicopter that he would build even three inches off the ground without breaking the law these days. His human powered helicopter design would be an unregistered and uncertified aircraft in the United States' air space once it was built and would be subject to confiscation and he would get six months in jail if he allowed myself to be caught flying his human powered helicopter in the United States off of private property. Which is where deregulation of human powered flight is adamant, and human powered helicopters should be seen and privileged as an environmentally non-invasive contrivance of aircraft capable of sustained flight, and not requiring a license to operate. In which case you will probably need a good life insurance policy because these are the beginnings or the 21st Century, not the 22nd Century, and human powered helicopters are just about to be born in to this world and one will not be the most technologically advanced new thing since the model-T 100 years ago was not a speedy Indy race car of the 21st Century, so give leeway to the effect that 100 years have to pass before there will be the most technologically advanced human powered helicopter the world has ever seen.

Up the chimney will be a game: fly to a high altitude and cease cranking, let the human powered helicopter plummet to the Earth, and with precise enough timing engage cranking just in time to slow to a stop and touch the ground with the landing gear. Children will get good at this game. It is like driving the tractor when cutting the grass, practice, practice, practice.

In the most probability the favorite altitude will probably be to fly just above the tree-tops. This will facilitate fewer injuries from falls should a human powered helicopter fail to perform for some reason. A fall in to the trees may cause less serious injuries, and falls from low altitude as well will cause fewer serious injuries.

Human powered helicopters may not be user serviceable except for greasing bearings, adding hydraulic fluid, making minor changes to the operator's compartment or changing a chain, and some more serious replacements of some parts. Most of the human powered helicopter is not user serviceable: however, the human powered helicopter is designed to be maintenance free and should give more than a lifetime of good service, with proper care.

Human powered helicopters also present the formidable capability of spying on groups or individuals. Going in seeking of "where it's at" is also a formidable prospect for the use of a human powered helicopter. Catching a boyfriend or girlfriend cheating will be a capability of human powered helicopters. The flocking together of individuals will occur. It will at once be possible to see soul mates from a distance in the air, and flocks of people may form in time. This will also facilitate that these individuals will accumulate in neighborhoods as companions, or extended "families". Extended families, akin to flocks of birds, would more probably be the wiser organization of flocks of people rather than organizations of large flocks of "friends". Flocks of "friends" would be more likely to be disorganized and leave a trail of infractions in their wake.

Once the neighborhood has been established, then it will be possible to design cities and towns or communities around the human powered helicopter, with spokes of taxiways radiating out from a central runway. Houses, homes, cities etc. may be built on these taxiways: however, if HTOL human powered aircraft are not employed then with landing platforms on the roofs of houses there may be no need for taxiways or a runway for human powered helicopters. It may even be possible to have a job in another city should the human powered flight prospect prove fast enough to get to work with if one lives in another town from their workplace. However, in the meantime parking will be at a premium for human powered helicopters in parking lots.

Celebrities will have to carry umbrellas if they don't want to be recognized from the air. Nudist colonies will have to cover up. One day there will be a law that you cannot fly a human powered helicopter unless one flies wearing no clothes, but then this may be in extreme circumstances of the necessity to have to re-populate the planet should the human population drop to, say, twelve individuals or some critical limit where people are dying faster than they are being born and the population will become extinct in a matter of, say, half a generation. No one can tell the true circumstances of the Revelation To John unless they are Saved by The Grace Of God and can truly interpret the scriptures with reality. Who knows? Maybe there will be a time in the future when the population of the Earth is in danger of becoming extinct in a short time. Until then, this wearing-no-clothes business is just foolishness for me. Perhaps it is true, but how can I tell the future?

Nevertheless, putting restrictions on human powered flight, certainly for human powered helicopters, will only embitter the population, like trying to make them stop drinking coffee, with some sort of jurisdictional restriction.

The human powered helicopters should come equipped with a GPS, an altimeter, a turn and bank indicator, and a compass, and a two-way radio all incorporated in to a 12 volt DC electrical system with an alternator for power and all the necessary implements of 12 volt DC electrical systems with all loads running in parallel. This should be sufficient equipment to operate a human powered helicopter in Federal airspace open to the public. Once again these human powered helicopters are designed to be pilot-license-free and deregulated so mankind can enjoy the freedom of the skies without the burden of regulations and be free as birds to fly in the common air, unhindered by any persistent influence of justice during their traverse.

Speaking of which, there is the invasion of privacy issue, should one appear in the sky where another is in anticipation of their privacy being kept confident: therefore, privacy (in Kentucky) is lawfully maintained at an arm's length-to be to the tips of one's fingers as far as one can reach, and should there be any dispute as to whether one is "invading one's privacy" by appearing somewhere over the horizon, it shall be noted that in Kentucky privacy is limited to an arm's length and should any dispute arise from a "spy" anticipating the capturing of views, photographs, or whatever kind of media a record can be made on, of some individual, group et cetera, so as to ascertain facts-on-the-ground that the limits of privacy do not extend "as far as one can see" but to the tips of ones fingers as far as one can reach. Therefore privacy does not extend in to space as far as one can see, nor does ones property line extend vertically to the outermost reaches, but it is only an arm's length to the tips of one's fingers as far as one can reach. One should then be able to come in to view, approach, land and walk up to the suspected subject, or suspect, and stay just outside the full reach of their fingertips and NOT BE INVADING THEIR PRIVACY and take pictures, make recordings, or just look at them, just to get the wisdom and understanding that one should choose otherwise another mate, gather information or whatever is required. As long as one does not come inside of the arm's length extenuation then privacy is not invaded. This arm's length law needs to be established throughout this country in the event of human powered helicopters and the capability of being able to ascertain wisdom and understanding in the light of one's relationship with another. Breaking off a relationship on the pretense that you succeeded at finding your mate cheating, because you have a human powered helicopter and it gave you the capability to spy on them, is every reason to establish an arm's length privacy law nationwide and to outlaw privacy being "as far as one can see" (in any direction), because the mate will use the defense "as far as I can tell, or as far as I can see, or as far as I can understand" or such like, to establish their privacy in the courts, and the judge will need to establish that the "as far as..."rule, anything when it comes to privacy is that the realm is limited to within the tips of one's fingers as far as one can reach: as well, preferential treatment (or mistreatment in the case of a male defense witness) for the female prosecution witness by judges, and the defense council, in courts needs to be outlawed with fines or imprisonment or both and should be investigated.

Vulcanization of property to human powered helicopters should be limited to any three points on the property, whether it be a building, growing structures, vehicles, fences, persons et cetera, and the lines that join the three points creating an assembly of areas covering the entire property inside the property line. And the consolidation of properties between neighbors to increase the size of the vulcanized perimeter should be restricted.

Considering the prospects of one stealing a human powered helicopter: once before from the previous design it was not possible because the device was constructed to tailored tolerances for each individual, but now the aircraft is sustainable in flight by articulation of the blade vanes and some variation in operators weights can be accommodated for. So it may be possible for someone to manage to perpetrate theft of a human powered helicopter and make off with one if their weight is not too significantly different from the owner's. The human powered helicopter will need some kind of lock.

This human powered helicopter would not be too purposeful for use in the military although it may be useful for moving troops in large numbers over great distances in a shorter period of time than presently accommodated for, it is vulnerable to attack since the powerplants are made of Styrofoam and epoxy-carbon fiber. The operator is exposed to raw gunfire and has virtually no protection besides body armor. This aircraft may prosper for military means some distance behind the front.

There is also the prospect from the Holy Bible that the high places shall be made low and the valleys shall be raised up even with the plains. This may be a reference to human powered flight.

As for farming, the human powered helicopter may be used to lift somewhat heavy objects and carry them distance, or do some heavy lifting of sorts. Also, as with minor heavy lifting the military or Department of Transportation may be able to use the human powered helicopter to go on search and rescue operations. The human powered helicopter may also be used as a stealth aircraft because it may be very quiet. The Corpman may have use of the human powered helicopter in combat, with the help of a GPS and the coordinates of injured soldiers.

Going camping will now be a strategy curse for the camper-outer. The police will be like Were-wolves on human powered helicopter campers because of their likelihood of carrying illegal drugs despite the fantastic corporate reality that human powered helicopters will cost probably $280,000 for a new one easily and that casual consumers of controlled substances won't be able to go human powered helicopter camping because they won't own one. But who will go human powered helicopter camping anyway? Human powered helicopters are designed with carrying capacity now and practical camping is within the means of human powered flight. So beware of the police cracking down on you for using drugs if you have a human powered helicopter. Be aware thou owner of thine human powered helicopter, thou art a drug addict, and the police are going to make you know this, so be prepared to be searched and taken into custody if you are brought down by a police officer, and to spend the night in jail and to have your human powered helicopter impounded or confiscated. Therefore, the Rebellion: The Rights of the People to Possess, Own, and Operate human powered helicopters in Federal airspace and between the ground and their surface ceiling without fear or interference from law enforcement in that an individual shall be free on their journey (this implicitly includes California, without exception and without exclusionary rules applying, California is to be disciplined in its mistreatment of interstate traffic and corrected). Also, The Smokey Mountains are to be managed by Federal Authorities in the careless mistreatment of interstate traffic and corrected where human powered flight is prejudged as a means of trafficking in controlled substances or simply as a means of transportation for a drug addict when an individual on a human powered helicopter is seeking to engage in wilderness adventure.

Any location of public interest, mainly national parks and forests, recreation areas et cetera, where individuals are found to be being mistreated having been perpetrated for offenders for being occupants of human powered helicopters will be investigated and the suspected department of law enforcement subject to disciplinary action and corrected if there were no charges for drug offenses and all the subjected offenses were a result of the officer(s) having apprehended the subject(s) and all the offenses were ex post facto, e.g. concealed deadly weapon, public intoxication, assaulting a police officer, resisting arrest, none of which are drug related, and all are after the fact, since the police office arriving on the scene is more than likely after an individual on a human powered helicopter for drugs than for a pocket knife, or for having had one beer, or for talking with their hands, or for struggling because the officer just broke their arm having had them in a hammer lock. Nevertheless, the fact remains that the motive for the mistreatment has to have been drugs related, and finding out whether or not there are any drugs during the scene during the arrest is paramount in the discovery of the motive during the investigation. No police officer shall be left unturned.

If you leave the human powered helicopter alone long enough spiders will make their webs in it, birds will make their nests in it, the paint will flake off, it'll rust, it'll rot, it will leak, it'll get wet in the rain, snow will collect on it and ice will hang down off of it. It'll collect dust. The battery will go dead, branches will fall on it. Mud will get splashed on it. Parts of it will fall off or get stolen. It will sink in to the earth and begin to lean over. Eventually it will fall over and no one will pick it up neither can it pick itself up, and after a long enough time it will turn to dust. It is not a God. It cannot see or hear or speak or work miracles. It attracts lightning it does not send it forth. It can be broken into pieces and smashed to smithereens. If you ask it a question it does not respond because it cannot answer. It is lifeless and does not even know what it is. It does not know anything. It is a mechanism, a machine, which man operates with his mind and body. It itself can do nothing. If something breaks on it while you are in flight you could die, it cannot save you it has no powers of self-preservation. Do not worship it. It is made by craftsmanship as are all machines. Human powered helicopters are not Gods, or God, or any God. They are machines and are to be cared for as machines and not neglected lest they fall in to disrepair for they age as mankind ages because of the imperfection of sin and their days are numbered. This human powered helicopter is however a manifestation of The Will of the Christian God (I am), Jesus the Son of God, and The Holy Spirit having used a mere human to work as His conduit for doing His Will. The Human Powered Helicopter however is not yet manifested in physical form, only in design concept reflecting that mankind is not ready for human powered flight at this time, but soon. The fact that the author is financially unable to accomplish the construction of a human powered helicopter leads him to believe that he has much research to perform still and many months of study as well because there are still lots of loose ends in the development of a practical human powered helicopter.

I hope you enjoyed this manuscript of how to get power from a bicycle. There are still many details which are not presented here: the precision alignment of parts, their close tolerances, how to contain the hydraulic fluid while it is under extreme suction force while it is not really but it really is (it's difficult to explain), shear force areas of impeller blade vanes and input and output shafts, there are so many calculations. The formula for fluid mechanics of inertia solves them all using the cubit and the fold and the base formula. Practice practice. Apply this formula to everything that moves and has force and there problems will solve. Good luck.

Richard Chastain

To Contact the Author Send Your E-Mail messages to:

Rick.l.chastain@gmail.com

Thank you for being an interested participant in the study and understanding of Volume 1 of the work in progress. Further development of these ideas and discoveries will prove to be a valuable experience. I'm glad you took the opportunity to challenge yourself with this volume of work.

Richard Chastain M.E.T.

Printed in the United States
By Bookmasters